Volume 1

PLACE AND POLITICS

PLACE AND POLITICS
The Geographical Mediation of State and Society

JOHN A. AGNEW

Routledge
Taylor & Francis Group

LONDON AND NEW YORK

First published in 1987

This edition first published in 2015
by Routledge
2 Park Square, Milton Park, Abingdon, Oxon, OX14 4RN

and by Routledge
711 Third Avenue, New York, NY 10017

Routledge is an imprint of the Taylor & Francis Group, an informa business

British Library Cataloguing in Publication Data
A catalogue record for this book is available from the British Library

ISBN: 978-1-138-80830-0 (Set)
eISBN: 978-1-315-74725-5 (Set)
ISBN: 978-1-138-79865-6 (Volume 1)
eISBN: 978-1-315-75658-5 (Volume 1)
Pb ISBN: 978-1-138-79866-3 (Volume 1)

Publisher's Note
The publisher has gone to great lengths to ensure the quality of this reprint but points out that some imperfections in the original copies may be apparent.

Disclaimer
The publisher has made every effort to trace copyright holders and would welcome correspondence from those they have been unable to trace.

MIX
Paper from responsible sources
FSC
www.fsc.org   FSC® C013604   Printed and bound by CPI Group (UK) Ltd, Croydon, CR0 4YY

PLACE AND POLITICS

The Geographical Mediation
of State and Society

JOHN A. AGNEW

Boston
ALLEN & UNWIN
London   Sydney   Wellington

Allen & Unwin, Inc.,
8 Winchester Place, Winchester, Mass. 01890, USA
the US company of
Unwin Hyman Ltd
PO Box 18, Park Lane, Hemel Hempstead, Herts HP2 4TE, UK
40 Museum Street, London WC1A 1LU, UK
37/39 Queen Elizabeth Street, London SE1 2QB, UK

Allen & Unwin (Australia) Ltd,
8 Napier Street, North Sydney, NSW 2060, Australia

Allen & Unwin (New Zealand) Ltd in association with the
Port Nicholson Press Ltd,
60 Cambridge Terrace, Wellington, New Zealand

First published in 1987

Library of Congress Cataloging in Publication Data

Agnew, John A.
  Place and politics

Bibliography= P.
Includes index.
1. Political Sociology. 2. Geography, Political.
3. Scotland--Politics and Government. 4. United States--
Politics and Government. I. Title
JA76.A36   1987   306'.2   86-32097
ISBN 0-04-320177-6 (alk. paper)

British Library Cataloguing in Publication Data

Agnew, John A.
  Place and politics: the geographical mediation of
  state and society.
1. Geography, Political
I. Title
320.1'2 JC319
ISBN 0-04-320177-6

Typeset in 10 on 12 point Bembo by
Computape (Pickering) Limited, North Yorkshire
and printed in Great Britain by Mackays of Chatham

"All politics is local"
  Thomas P. "Tip" O'Neill
  Speaker, US House of Representatives (1984)

"We speak a different language and here
our tongue is incomprehensible"
  Carlo Levi, *Christ Stopped at Eboli* (1947)

"Signs and the signs of signs are used
only when we are lacking things"
  Umberto Eco, *The Name of the Rose* (1983)

Preface

Pistoia and Lucca are two cities in central Italy that are not very far apart, about forty-three kilometers, but have completely different political profiles. Pistoia is and has been for many years a stronghold of the Italian Communist party, Lucca is a Christian Democratic city in a region in which the party of that name is a minority one. Outwardly the two cities are very much alike – for example, they have similar demographic and social characteristics. But their histories as *places* have been quite different. In the case of Pistoia the city has long been "controlled" from outside, principally from Florence, and the surrounding countryside is characterized by an intense anti-clericalism that survives from the time when the Church was the major local landowner. In the case of Lucca, the city was the seat of a long independent city-republic that persisted until the late eighteenth century and a countryside in which peasant proprietorship was long established. Political life in these two cities, therefore, can *only* be understood by reference to their development as places. The central theme of this book is that this is equally the case elsewhere.

But a major contention of modern social science is that "time conquers space" or that, as history proceeds, differences between places become less and less relevant to social life and hence should have less and less of a role in social theory. This book disputes this contention in the context of one important aspect of social life: political behavior.

The frontispiece quotations capture metaphorically the three major contentions of the book: first, that political activity is structured and realized through place-specific social processes (O'Neill); secondly, that social science in general and political sociology in particular have not paid sufficient attention to this (Levi); preferring instead, thirdly, to organize explanations of political activity entirely through the inferred political dispositions of social categories (or individuals as members of them) rather than the *causes* of political activities in places (Eco). To the extent that a geographical concept of place has survived in recent social science, it is in terms of the geographical impacts of social processes operating through social categories. Places in themselves are rarely seen as *constitutive* of social behavior.

Place is a complex word. The Oxford English Dictionary gives over three and a half pages to it. Symbolic of its devaluation as a geographical term meaning "a portion of space in which people dwell together" or "locality," place is now quite often a temporal or social-categorical term meaning "rank" in a list ("in the first place"), temporal ordering ("took place"), or "position" in a social order. In social science, the emphasis that a *geographical* concept of place puts on social context or milieu has been largely eschewed in favor of an emphasis on either social categories (classes, status groups) defined by compositional analysis of national populations or the placeless "individual". Ironically, therefore, place *has* sometimes appeared in social science but as a social-categorical rather than a geographical concept!

The fundamental question is how to keep a focus on the individuality of places while acknowledging the real *interdependence* of places to which social-categorical analysis has attempted to respond. The issue is important. It is often viewed in terms of the relation between the unique and the general. Marx, obviously someone who had no aversion to the general, nevertheless captured the issue when he wrote in a letter in 1877:

> Events strikingly similar, but occurring in a different historical milieu, lead to completely dissimilar results. By studying each of these evolutions separately and then comparing them, it is easy to find the key to the understanding of this phenomenon; but it is never possible to arrive at this understanding by using the *passe-partout* of some historical-philosophical theory whose great virtue is to stand above history (quoted by Carr 1961:82).

But the issue of place individuality in the face of place interdependence also involves the question of the *mediation* between state and society. Territorial states are made up of places in which the social relations upon which social and political order depend are constituted. Places therefore do not exist independently, but as *sets* of places interrelated by state-building and associated social practices (MacLaughlin & Agnew 1986). State and society are not autonomous and separable therefore. It is a fallacy to separate them conceptually as sharply as we often do (Desai 1986). One reason we can do so is because at the same time we separate them analytically, we usually empirically confound them. The 'mediation' of state and society has been accomplished by merging them. Of special importance, the social categories of the state census have come to serve as the major mediating concepts of empirical social–political analysis. There has been no 'conceptual space' for the mediating role of place. The 'Jacobinism' of much social science has led to an 'official silence' (Perrot 1984:54) about the role of locality or place in mediating the actual interrelationships of state institutions and social processes.

This book originated somewhat parochially as an attempt to explain the successes of the Scottish National party in the 1970s. Existing accounts based largely on social-categorical analysis were not convincing. They could neither explain who was strongly attracted to the party nor why. This was joined by a growing dissatisfaction with conventional approaches to explaining political behavior. In a paper written with James S. Duncan (Agnew & Duncan 1981), I had examined some of the presuppositions or antinomies underlying contemporary social thought and found them wanting. Conventional political sociology seemed to reproduce these pre-suppositions without giving much attention to them. I therefore decided that my unease was as much philosophical as naively empirical.

The content and organization of the book represent the dual origins of my interest in place and politics. The early chapters focus on some major conceptual questions posed by a reading of contemporary political sociology and the answers provided by an alternative perspective. The later chapters take up certain empirical themes, in particular as expressed through my interest in Scottish politics and as expanded to an examination of American politics. Place is important in the 'New World' as well as in the 'Old'.

The project only came to fruition after a longer-than-usual visit to Scotland in the summer of 1983. This led me to ask anew questions about place and politics I had first raised ten years previously. It also coincided with the publication of a number of books, such as Rokkan and Urwin (1983), Burnham (1982), and Bensel (1984), that *pointed* in a similar direction.

This book appears at a time of revitalized interest in achieving what could be called "contextual" social theory. It draws explicitly on the pioneering theoretical work of Giddens (1979, 1981), Foucault (1980) and Pred (1984). But it also draws on older sources in the geographical literature, such as George (1966), and on the Gramscian perspective on the state. Foucault (1980:149) encapsulates what is most at stake for this diverse literature when he writes:

> Space was treated as the dead, the fixed, the undialectical, the immobile. Time, on the contrary, was richness, fecundity, life, dialectic ... The use of spatial terms seems to have the air of an anti-history. If one started to talk in terms of space that meant that one was hostile to time. It meant, as the fools say, that one "denied history" ... They didn't understand that [these spatial configurations] ... meant the throwing into relief of processes – historical ones, needless to say – of power.

In writing a book numerous debts are incurred. I am especially grateful to the Maxwell School at Syracuse University for tolerating my unconventional social science and supporting my extended visit to Scotland in 1983 through an Appleby–Mosher Fellowship. During my visit John Bochel of the University of Dundee was a gracious host and vital informant. Paul Knox, then of Dundee and now of Virginia Tech., Jack Brand of the University of Strathclyde, Allan Macartney of the Open University, and Henry Drucker and Hugh Bochel of the University of Edinburgh also gave me important information and useful comments on contemporary Scottish politics. Gordon Wilson, MP, and Chris Maclean of SNP headquarters in Edinburgh gave freely of their time to discuss the problems and prospects of the Scottish National party. A number of people have read portions of the book or an entire previous draft and offered useful if often skeptical comments. Special thanks are due to Tom Bottomore (University of

Sussex), Derek Urwin (University of Warwick), Gordon Clark (Carnegie-Mellon University), Peter Jackson (University College, London), and David Ley (University of British Columbia). Though grateful for their comments and suggestions I absolve them of any responsibility for what is contained in the final version. I am also grateful to Harriet Hanlon for typing the manuscript and Marcia Harrington for drawing the figures. Ellen R. White of Michigan State University prepared Figure 10.3. Finally, I would like to thank my family in Scotland for putting me up in 1983 and my family in Syracuse – Susan, Katie, and Christine – for putting up with me during the writing of the book in the years since.

John A. Agnew
Syracuse, NY

Acknowledgements

I would like to thank the following individuals and organizations who have given permission for the reproduction of text figures and tables.

Figures: Sage Publications (6.1, 6.10); The American Enterprise Institute for Public Policy Research (6.2, 6.3); The American Political Science Association (6.5, 6.6); Oxford University Press (6.7); Macmillan and St Martin's Press, USA (6.8, 6.9); Batsford (7.1); Royal Dutch Geographical Society (7.2); University of Wisconsin Press (10.1, 10.2); Association of American Geographers (10.3).

Tables: Macmillan and St Martin's Press, USA (6.1); University of Wisconsin Press (10.1, 10.2); Pantheon Books (12.1); Oxford University Press (12.2); Table 12.3 from *Election '84: landslide without a mandate*, edited by Ellis Sandoz and Cecil V. Crabb Jr; copyright © 1985 by Ellis Sandoz and Cecil V. Crabb Jr, reprinted by arrangement with the New American Library, New York.

Contents

List of tables

1 Introduction

There is both an academic and a popular demand for simple answers to large problems. Intellectuals have acquired much of their influence from their ability to supply formulae that do just that. They have, so to speak, provided a "glue" that makes impressions of the world cohere. Their explanations not only enable people to make sense of the world around them but also determine in large part *what* they see around them. The field of political sociology – generally, the study of the social bases and background to political activities – has produced a number of key formulae that have become institutionalized over the past forty years. These share common assumptions, even if they also differ in various ways (see Ch. 2). A critical one is a unitary vision of society defined by the boundaries of the state. Explanations of political activities are sought in categories and concepts that relate all social cleavages to the level of the state. Febvre's dictum *"L'État n'est jamais donné, il est toujours forgé"* (Febvre 1922) has been forgotten.

This book is an attempt to move beyond political sociology as defined by "the cult of the state" and focus on the geographical rootedness of political life. The central premise is that territorial states are made out of places. This is not to argue that state institutions are not implicated in political life. Far from it, they have become preeminent. Rather, it is to argue that the social bases of response and resistance to these institutions are best viewed in terms of the histories of places rather than in terms of universal or national structures (such as class-as-a-category) or individual agency (the isolated individual's response to a national political stimulus).

The persistence of place-specific and regional voting patterns provides some support for the importance of place as a *correlate* of political activities. Of course, whether place is of *explanatory* significance remains at issue. The modernization theories that have dominated American social science since World War II (Tipps 1973) have had no role for place. Amongst popular *genres* of social science only the *Annales* school of social history and the idiographic, and largely atheoretical, tradition in historical geography have maintained a stress on place in explaining social and political behaviour. Sometimes this has reflected a backwards-looking or antiquarian orientation. Classical French social geography and the political geography of Siegfried (1913) both combined a focus on regional patterns with a lack of interest in regional *change* or social causation. Historical geography has largely continued in this vein. Focusing on places in the past or the "yoke of the past" in the present, they have tended to ignore questions of social and political dynamics. Only recently have exceptions to this rule appeared (e.g.,

Gregory 1982). But the *Annales* school, if "school" is still an appropriate term (Aymard 1972), has been at the center of attempts to map and *explain* the complex reality of human life by means of local and regional studies (Baker 1984). Its hallmark has been a priority to causal validity *in* places rather than generalization *across* places. There has also been an imperative to *push* local history through to the present; to understand the ways in which contemporary local *mentalités*, practices, and social structures have emerged out of earlier ones (Goubert 1971, Leuilliot 1974, Thuillier 1974). This emphasis provides an important inspiration to the perspective of this book.

Often in social science, places are merely the testing grounds for concepts or hypotheses with presumed general or universal significance. Such an approach denies the *distinctive* social and historical characteristics of places. They are case studies or instances of general "laws" or "tendencies." An interesting example of this use of places would be Foster's *Class struggle in the Industrial Revolution* (1975). In his study the places selected are mere *instances* of a general process. By way of contrast J. Smith's (1984) "Labour tradition in Glasgow and Liverpool," a paper referred to in Chapter 8, treats the places investigated as historically evolving and distinctive milieux rather than case studies of some overarching and transcendental historical dynamic.

Though not denying that human behaviour is subject to statistical *description*, it is simply inadequate to stop at that point if the objective is *explanation* (Szymanski & Agnew 1981). Any attempt at explanation in social science is faced with the need to identify the causes of human behavior. A major problem has been how to do this while allowing for the manifest reality of human agency. This in turn has directed attention to places as the social contexts for human behavior – the settings in which activity is caused and takes on meaning and purpose (Pred 1984). Thus, places, rather than being particular instances of general laws, are made out of human practices.

Even in a world dominated by a global division of labor, place maintains its significance. As N. Smith (1984) points out, uneven geographical development is a marked and continuing feature of the modern world. This reflects a perennial search by businesses for higher rates of return on capital invested and consequent disinvestment elsewhere on the part of global- and national-oriented business enterprises. The modern world-economy is differentiating in its effects more than equalizing. But place also retains importance because of the differential *reaction* of people in different settings to uneven development (C. Smith 1984). To insist on the continuing importance of place, therefore, is not to deny that processes beyond the locality have become important determinants of what happens in places. But it is still in places that lives are lived, economic and symbolic interests are defined, information from local and extra-local sources is interpreted and takes on meaning, and political discussions are carried on. Even in a world in which many sources of information and social cues are extra-local, especially as transmitted by the electronic media, information and social cues are meaningful only when activated in everyday routine social interaction. For most people, as later chapters should make clear, this is still defined by the locality. Home, work, church, school, and the like still form nodes around which everyday life circulates (Pred 1984).

In modernization theories, place is viewed as significant only in traditional or parochial societies and not in "modern" ones. These are "non-place realms" (Webber 1964) in which only "national" societies, as defined by modern armies and statesmen, are important and real. An extreme is reached in those accounts that see identity entirely in terms of class or other categorical abstractions *divorced* from the practices of everyday life. Two major political sociologists, Lipset and Rokkan (1967 : 18–19), for example, contend that:

> The National Revolution forced ever-widening circles of the territorial population to choose sides in conflicts over values and cultural identities. The Industrial Revolution also triggered a variety of cultural counter-movements, but in the longer run tended to cut across the value communities within the nation and to force the enfranchised citizenry to choose sides in terms of their economic interest, their shares in the increased wealth generated through the spread of new technologies and the widening markets.

"Nationalization" and the declining significance of place or locality are likewise major themes in the political science and sociological literatures on "nation building" and the effects of national mass media on political behavior (e.g., Deutsch 1953, Geertz 1963).

At the outset, it is important to stress that the "nationalization thesis" is not just an intellectual position. It is also a political one. Therefore in criticizing the nationalization thesis one inevitably takes an antithetical political position. Its view of social change as the dissolution of customary, small-scale social life and the concomitant growth of national "citizenship" is familiar to us all from our school days. The implication is that the state, engaged, as Tilly (1979 : 19) puts it, in "a *mission civilisatrice*," broke down the barriers of "local barbarism." Light overcame darkness. But to argue to the contrary is not just to argue for place, so to speak, it is also to argue against this benign view of the state. Rather than a shift from autarky to control, state-building involved a shift in control over local interests from local populations and elites to national capital and the national state. However, this shift has been incomplete. Even in so-called totalitarian states such as Nazi Germany, state control was far from total, even if much greater than in most liberal democratic states (Kershaw 1983). In conditions of economic and political crisis, resistance, in the form of sectionalism, place-based mass abstention from established political routines, or political violence directed at the state, can threaten the continued existence of the state itself or the form that it presently takes.

A recent explicit effort to demote the importance of place argues that the electronic media have effectively separated "social place" from "physical place" (Meyrowitz 1985). Information saturation across physical places has created an increasingly undifferentiated social terrain. But even Meyrowitz acknowledges the limits of his argument. His last example, concerning airport lounges and the constraints they impose on social interaction, is of a physical place. Moreover, he provides a number of cases where the media are a "double-edged sword" as far as physical places are concerned. For example, many media-favored social groupings (Meyrowitz mentions "smokers" and "nonsmokers") are based on one superficial attribute of people rather than complex and long-term *shared* experiences. The banality of such designations is apparent to all but the most incredulous. More importantly, in finding out more about other places and their inhabitants, one's differences with them may be *enhanced* rather than diminished. At one scale, nationalism, and at another scale, parochialism, feed off the stereo-types in which the electronic media tend to trade. They reinforce rather than undermine the identity between, to use Meyrowitz's terminology, "social places" and "physical places."

To the extent that geographical variation in social organization and political behavior has been taken seriously in recent social science and political sociology, it has taken the form of either a focus on the spatial distribution of individuals exhibiting different traits and fitting into prede-fined national census categories *or* an emphasis on evidence for sectionalist or separatist political tendencies in specific regions. In the first case, geo-graphical differences in levels of support for political parties or political participation in general are viewed as "composition effects." Geography is epiphenomenal, it is merely the aggregate product of "individual" attributes that just happen to covary with location. For example, more poor people live at some locations, more affluent people live elsewhere. The second views territorial or sectional interests as capable of overriding national cleavages in certain circumstances to produce a geographically differentiated pattern of political expression. In particular, "core–periphery" cleavages in economic performance and political power can create conditions to which peripheral sectionalism (and core sectionalism?) is a necessary response. This has been the preferred alternative in attempts to account for the recent growth of separatist movements in Western Europe and Canada and other evidence against the nationalization thesis (e.g., Hechter 1975, Nairn 1977, Gottman 1980, Archer & Taylor 1981, Bensel 1984). In both cases, however, geography is *extrinsic* rather than *intrinsic* to political behavior. Ultimately, variation from the national "norm" is due to "special factors" that do not presuppose the inadequacy of conventional accounts in other settings and under other conditions.

Surprisingly, perhaps, geographers have usually shared the dominant models of spatial variation with other social scientists. In the realm of political sociology, electoral geographers have concerned themselves with the spatial organization of electoral areas, boundary definition, gerrymandering, and so forth; mapping election results; and explaining election results in terms of certain "behavioral processes" such as interpersonal information transmission, reception, processing, and electoral choice. By and large they have drawn their inspiration from the literature of political sociology and share its biases (see, e.g., Johnston 1979, Taylor & Johnston 1979). One of these is a penchant for juxtaposing within one study a macroscale approach to the state, such as that of Easton's (1953) systems analysis, with a microscale or behavioral approach to political behavior (Johnston 1979, Muir & Paddison 1981). Another has been a tendency to avoid the concept of place or see it entirely as a source of "local effects" that can be divided from the effects of other scales in explaining political behavior (e.g., Johnston 1976). Geography in the 1960s and 1970s, when electoral geography became popular, was, as a discipline, in retreat from its "exceptionalist" tradition. This tradition, it was argued, generated a stress on "areal differentiation at the expense of areal integration" (Haggett 1965:3). Geographers told one another that the scientific status of their discipline depended upon the acquisition of generalizations. Such generalizations, it was hoped, would come if geographers studied spatial form and patterns. Insofar as processes are invoked, they tend to refer to "factors" affecting spatial form rather than to the geographical imprint on social process implied by the concept of place. Recent geography therefore is not much help in developing a *geographical* political sociology.

It is in fact to recent developments in sociology and to the work of a small coterie of geographers that one must turn for conceptual materials that provide the building-blocks of a place-based political sociology. In particular, the writings of Foucault (1980), Giddens (1979, 1981), and Pred (1983, 1984) are all sensitive to the fact that human activities take the form of concrete interactions in time–space. They all argue, if in different ways, that in order to explain human behavior one must deal with the material continuity of everyday life, or the process of "structuration" whereby the *structural* properties of social life are expressed through everyday practices which in turn produce and reproduce the micro- and macrolevel structural properties of the social groups in question (Pred 1983). Attention is thus directed to the settings and scenes of everyday life: to place.

In a more analytic vein, three aspects to place can be identified: locale, location, and sense of place. *Locale* refers to the structured "microsociological" content of place, the settings for everyday, routine social interaction provided in a place. *Location* refers to the representation in local social interaction of ideas and practices derived from the relationship between places. In other words, location represents the impact of the "macro-order" in a place (uneven economic development, the uneven effects of government policy, segregation of social groups, etc.). *Sense of place* refers to the subjective orientation that can be engendered by living in a place. This is the geosociological definition of self or identity produced by a place. These three elements of place are discussed in more detail in Chapter 3.

The major theme of this book is easily stated. It is to argue that political behavior is intrinsically geographical. The focus is on the concept of place. The social contexts provided by local territorial–cultural settings (neighborhoods, towns, cities, small rural areas) are viewed as crucial in defining distinctive political identities and subsequent political activities – from votes to strikes to street violence. In this book, votes receive rather more emphasis than other activities only because they have become the currency of political sociology rather than because they are more "special" or necessarily more legitimate than other activities.

The book is organized into 12 chapters. Chapter 2 provides a review and critique of dominant modes of theorizing in political sociology. This is not a complete review but an attempt at identifying an intellectual crisis in the existing literature to which this book is responding. Chapter 3 lays out the argument for explaining political behavior in terms of place. Chapter 4 examines some important themes in political sociology from the place perspective. Chapter 5 attempts to explain why the concept of place has not received much attention or welcome in social science and political sociology. Chapter 6 provides empirical evidence questioning the "nationalization thesis" that lies at the heart of most contemporary political sociology, and critically reviews various attempts to deal with this. It ends by proposing the place perspective as a superior alternative. Chapters 7 through 12 present attempts to illustrate and support the place perspective using the examples of popular political behavior in Scotland (Chs. 7 to 9) and the US (Chs. 10 to 12) over the past 100 years. Taken together it is hoped that this book demonstrates the truth in the words that for political behavior, "It is the local reality that determines the total picture, and not the reverse" (Granata 1980:512).

2 Dominant modes of political sociology

The field of political sociology is characterized by several distinct modes of theorizing. One major division is normative; between those who are mainly preoccupied with the operations of existing political institutions and their role in creating political and social "stability"; and those who are concerned mainly with the forces which produce instability and possibilities for change (Bottomore 1979: 12). Other differences revolve around the presuppositions or assumptions about the nature of "man," "society," and "knowing" that underpin different *genres* of political sociology.

A major objective of this chapter is to describe the main or dominant modes of political sociology. Another objective is to pinpoint areas where the dominant modes are under challenge. Rather than provide a fully fledged review of political sociology today, the overall purpose is to provide a background for the sense of dissatisfaction with the intellectual status quo, indeed a sense of "crisis" in the field, that has stimulated interest in developing the alternative "place perspective" described in Chapter 3.

This chapter is organized as follows: first, attention is directed towards defining the term "political sociology"; secondly, the underlying presuppositions of dominant modes of theorizing are discussed, using examples of influential studies; thirdly, "areas of challenge" are identified; fourthly, and finally, the general condition of contemporary political sociology is assessed.

Defining political sociology

As an area of study political sociology can be defined both abstractly and in its particular manifestations. Abstractly, political sociology is concerned with "the phenomenon of power at the level of an inclusive society (whether that society be a tribe, a nation state, an empire, or some other type); the relations between such societies; and the social movements, organizations and institutions which are directly involved in the determination of such power" (Bottomore 1979:7). The study of such matters is ancient. However, the term "political sociology" is of recent vintage (Runciman 1969: 22). In its modern usage, definition is usually much narrower than Bottomore's abstract one. To Lipset and Rokkan (1967: 1–2), for example, political sociology is concerned with three sets of questions: "The genesis of the system of contrasts and cleavages within a national community," "the

conditions for the development of a stable system of cleavage and oppo-
sitions in national political life," and "the behavior of the mass of rank-and-
file citizens." These sets of questions form the core of *orthodox* or conventional political sociology today. But the focus is above all on the social conditions for political stability in national states (Kornhauser 1959, Lipset 1960,
Huntington 1968).

Such ideas lost some of their appeal in the wake of the acute political conflicts in Western industrial "societies" in the late 1960s. As a result there has
been a renewal of interest in an alternative or *antithetical* political sociology
which takes as its starting point the strains and conflicts in all societies, and
regards stability as only a temporary (and partial) resolution of deep-seated
antagonisms. This position, largely Marxist in inspiration, assigns a larger
role to the use of force rather than "value consensus" in producing and
reproducing a particular society, and views values themselves as the product
of coercive institutions such as schools, the mass media, and state agencies.
Furthermore, political institutions – party systems, types of government –
are seen as largely dependent upon underlying economic forces rather than
autonomous in their operation and effects (Bottomore 1979: 12–14).

These basic normative orientations aside, orthodox and antithetical political sociology share a number of features. One of these is their operational
focus on national-scale societies. Another is a commitment to an evolutionary conception of social change. Each accepts the view that as societies
"modernize" they undergo certain determinate mutations that create new
conditions for political life. In particular, national *functional* cleavages (such
as class) displace older territorial and communal cleavages (such as region
and religion). A final communality is their identification of power and/or
force with the modern national state. This tendency reifies power as a monolithic and repressive entity operating at a macroscale. It consequently
removes the possibility of seeing power in terms of creativity and agency in
microscale situations (Harris 1957, Foucault 1975, 1976).

It is possible to derive from all this something of a lowest common
denominator definition for contemporary political sociology. The subject
matter of political sociology, both orthodox and antithetical, is those institutions and behavior that are conventionally labelled "political" (Runciman
1969: 41). Today this means institutions associated with national states, and
behavior relating to attempts to share power over or to influence the distribution of power among social groups within these institutions.

Dominant modes of theorizing

PRESUPPOSITIONS
All students of human activities, whether they are explicitly interested in
explanation or not, necessarily adopt one of several "modes of theorizing."

These have three elements to them: a model of man, a model of society, and a model of knowing (Hollis 1977, Lukes 1977, Claval 1984). The researcher may or may not be aware of the mode that has been adopted. But it is impossible to avoid taking a position on the "images" of humanity, society, and knowledge that are *presupposed* by particular approaches. The three models that constitute modes of theorizing in contemporary social science are best expressed in terms of the paired opposites or presuppositions on which they are based: (*a*) model of man: voluntarism–determinism; (*b*) model of society: individualism–holism; and (*c*) model of knowing: positivism–intuitionism (Runciman 1969: 1–21, Ryan 1970, O'Neill 1973, Agnew & Duncan 1981). After discussing each of these, attention is directed to how they have been combined to form the dominant modes of theorizing in political sociology.

The conflict between voluntarism and determinism is an age-old dispute between theoretical frameworks, or problematics, over whether or not human behavior is willed or determined. Lukes (1977: 13–29) formulates the issue as that of the relation between power and structure. He identifies three positions that have been adopted. The first is the *voluntarist* position. From this view, the limits facing "choice-making agents" are minimal; in particular, the only "structural constraints" are *external* to the agent. Existentialism is an example of this position. Sartre (1959) in his *L'existentialisme est un humanisme* argued that man is "what he conceives himself to be ... what he wills ... Man is nothing else but that which he makes himself." He is "not found ready-made; he makes himself by the choice of his morality and the pressure of circumstances is such that he cannot fail to choose one" (Sartre 1959: 22, 78).

The second position is the *structuralist* position. This is most clearly exemplified in recent social science by the work of Althusser and his disciples. Althusser's (1970: 80) model of man is described well in the following passage:

> the structure of the relations of production determines the *places* and *functions* occupied and adopted by the agents of production, who are never anything more than the occupants of these places, insofar as they are the 'supports' (*Träger*) of these functions. The true 'subjects' ... are therefore not these occupants or functionaries, are not, despite all appearances, the "obviousness" of the "given" of naive anthropology, "concrete individuals," "real men" – but the definition and distribution of these places and functions. The true 'subjects' are these definers and distributors: the relations of production (and political and ideological and social relations).

At its most extreme, then, the structuralist position is that different levels of structural constraint (economic, political, and ideological) are determining of 'actual' behavior. Thus, power is an *effect* of "an ensemble of structures"

(Poulantzas 1973:101) rather than a *capability* to act on the part of human agents.

The third position is what Lukes (1977: 17) calls the *relativist* position. This holds that there are just "different points of view, or levels of analysis or problematics, and there is no way to decide between them." One can *choose* to adopt a voluntarist position, viewing individuals and collectivities as always exercising reason and will, or one can *take* a structuralist position, seeing individuals and collectivities as "wholly determined, acting out roles, and indeed being not merely influenced but actually constituted by ever pre-given structures of a system that operates upon them and through them" (Lukes 1977: 17–18). There is no way of choosing between them because the accounts of power and structure they offer are incommensurable (e.g., Winch 1963).

Another fundamental dispute in social science concerns the "model of society" that should be adopted. This is not independent of controversy over the "model of man"; they are related, but the dispute over the model of society involves the question of "what exists?" rather than "what determines?" Basically the question is this: when we speak of society, a social group, an economy, or a nation-state, what do we mean? The *individualist* maintains that statements about any collectivity must be reducible to statements about the individuals of whom the collectivity is composed (Runciman 1969:6–7). But the *holist* claims that there is a need for concepts "which depend for their meaning precisely on the fact that they can never be reduced to a list of assertions about individuals" (Runciman 1969:6). Some social scientists use holism in a much weaker sense to refer to a sensitivity to "context" or "relatedness" (e.g., Brookfield 1978, Ley 1978). But this is potentially misleading. The term holism has a well-established meaning: "wholes" (supraindividual entities) are greater than the sum of their "parts" (individuals) (Phillips 1976, Ryan 1970: ch. 8).

Individualism and holism have not enjoyed equal support at all times in modern intellectual history (Lukes 1973). Social and intellectual changes have often favored one answer to the question of "what exists?" over the other. Seventeenth- and eighteenth-century European science ushered in notions of individual freedom and autonomy and the doctrine of natural rights. Aristotle's theory of the state as "a product of nature" became largely unacceptable during the 17th century because it appeared to deny central ontological status to "the individual." It was supplanted by a contract theory of the state based on the voluntary actions of free individuals: religious and moral individualism, belief in the sacredness of individual conscience, and the "rights" of free opinion, free movement, and free worship were characteristic of this period in northern Europe (Macpherson 1962). However, early in the 19th century, idealist philosophy, especially that of Hegel and the German Romantics, brought about a resurgence in holistic thinking. In this century social science has been a scene of struggle between individualism and holism (Phillips 1976). The new "disciplines" of social science divided up largely in terms of their position in the individualism–holism controversy. Thus sociology and anthropology have been largely holistic and psychology and economics have been overwhelmingly individualistic in their ontological commitments.

The final model at issue is the "model of knowing." Again there are two polar positions. The two sides this time are the *positivist* and the *intuitionist*. The positivist position is usually expressed as the belief in the methodological complementarity of the natural and the social sciences (Bottomore 1979:14, Runciman 1969:8–13), in particular, the search for generalization or empirical laws. But as Keat (1971) has pointed out, as it stands this will not do. In contemporary social science, positivism refers to a specific conception of science, one that focuses on a particular set of procedures or instruments as the key to obtaining scientific knowledge (MacKenzie 1977). Natural science, however, is not practiced this way (Harré 1970, Keat 1971). But the positivist *image* of natural science is widespread in social science, and thus does involve the projection of *a* model of knowing from one domain, even though *not* characteristic of it, to another.

The intuitionist accepts the characterization of natural science in positivist terms but rejects the claim that the natural sciences are methodologically equivalent to the social sciences. To the intuitionist social science is not about discovering the causes of human behavior through experimentation or statistical analyis. It is concerned with studying human society by "understanding the meaning of intentional rule-governed action" (Bottomore 1979:14). Collingwood (1948) – one of the best known, if most extreme, intuitionists – argued that human behavior is not just "behavior" but has meaning for the humans performing it. Consequently, a distinctively *social* science is only possible if the "inside" of human *action* (reasons) is captured. Concern with the "external regularities" of observable behavior (and causes) is not enough (Collingwood 1948:214).

Intuitionism also stresses the "uniqueness" of historical events. Generalization such as that practiced by the positivist is irrelevant because historical events are by definition not replicable (Louch 1966). Thus, as Runciman (1969:9) puts it: "if the particular event has not been explained, it cannot be incorporated into a generalization, whereas if it has been explained, the generalization is unnecessary to explain it."

These paired opposites or presuppositions clearly represent extremes. In practice much social science *tends* towards, rather than locates at, a specific extreme. The point here is not to describe existing social science as much as to provide a conceptual grid for defining modes of theorizing in political sociology.

MODES OF THEORIZING

Taken together, these three sets of presuppositions provide a framework for categorizing modes of theorizing in political sociology. They combine in predictable – and unpredictable – ways. In contemporary political sociology there are four major varieties: (*a*) behavioral (voluntarist–individualist–positivist); (*b*) systemic (voluntarist–holist–positivist); (*c*) Marxist–voluntarist (voluntarist–holist–positivist); and (*d*) Marxist–structuralist (structuralist–holist–positivist). The first two constitute, in uneasy alliance, "orthodox" political sociology. The second two form, equally uneasily, "antithetical" political sociology. The second and third modes of theorizing seem the most eclectic and least internally coherent. In common parlance voluntarism and individualism are often equated, but this is mistaken. A voluntarist position can take a collectivist form when the units exercising will are groups rather than individuals (Lukes 1977:16). This is then *sociological*, as opposed to *psychological*, voluntarism.

The behavioral approach   Behavioral political sociology covers the bulk of recent work in voting and party research, much work on new political movements, and research on lobbying and interest group politics. There are, of course, important divisions within this literature, in addition to differences in subject matter. For example, two paradigms now dominate American voting research. One, the rational choice, grew out of Downs's *An economic theory of democracy* (1957). The other stems from the political psychology approach pioneered by Campbell *et al.*'s *The American voter* (1960). The rational choice proponents have developed rigorous logical models of individual decision-making and party behavior which have been resistant to empirical analysis. The party identification approach (the Michigan School) avoids deductive theorizing in favor of accumulating empirical findings, primarily from a succession of biennial American national election surveys emphasizing the relative impact on voting behavior of long-term "cues" (social attributes and party identification) and short-term "cues" (candidate perceptions and issue preferences). But these "paradigms" share most major presuppositions. They differ mainly in terms of emphasis on modelling as opposed to inductive inference (Budge *et al.* 1976).

Voting research became more behavioral in nature in the 1950s, but the antecedents of this now dominant mode of theorizing can be traced back to the 1920s. A series of conferences at the University of Chicago in the summers of 1923, 1924, and 1925 proposed a new "science of politics" (Jensen 1969a,b). The 1924 conference was particularly important. It introduced contemporary behavioristic psychology into political science. The psychologist Thurstone proposed that the concept of attitude should be adopted as the basic unit of "scientific" analysis in political science (Jensen 1969b). The potential for grounding the study of politics in the concepts and methods of psychology became widely accepted.

In the 1940s and 1950s, voting research merged this predilection for psychological measurement with a focus on the decision-making *process* which precedes voting. Two studies in particular developed a focus on the flow of information during campaigns as a major determinant of voting behavior. These were Lazarsfeld *et al.*'s *The people's choice* (1948) and Berelson *et al.*'s *Voting* (1954). But the Michigan School, wary of the sociological implications of these studies and their focus on specific places, focused on attitudinal variables and the political content and implications of voting (Sheingold 1973:713). They thus returned, albeit in a more methodologically sophisticated guise, to the program of the 1920s. Survey-defined "variables" are regarded as "brute data" with an invariant meaning across social situations and immune to questions about their status or validity that are not themselves empirically derived (Taylor 1971). More recently, attempts have been made to combine the findings of voting surveys with larger-scale analyses of electoral change (e.g., Campbell *et al.* 1966) but the basis remains the focus on the individual voter (Eulau 1976).

A different approach to voting research views the act of voting in terms of what it says about group affiliations within a society (Lipset 1960). This more "sociological" conception of voting is particularly concerned with changing "cleavage structures" and the shift from more "traditional" parochial loyalties to more "modern" national ones (e.g., Alford 1963, Duverger 1959, Blondel 1963). This tradition, though a "shade" more holistic in appearance than others, has in fact a nominalistic conception of such concepts as class or religion. Individuals "belong" to different social categories. These categories do not have an independent existence. Thus, even the apparently more sociological behavioral studies are voluntarist, individualist, and positivist.

Several features that pervade behavioral political sociology follow from the emphasis on surveys and acceptance of individuals as the sole units of analysis. First of all, voting lends itself to this approach. As a consequence other types of political activity, such as party activism, issue-related politics, and violence, are ignored or given short shrift because they are not amenable to positivist treatment. Secondly, and most importantly perhaps, behavior is divorced from institutions, and viewed as constituted by a simple feedback of actor response to political stimulus. Only the most wildly optimistic Chicago conferees in the 1920s could have expected that the proposals for a new "science of politics" they heard would have survived intact for the next sixty years without *any* major innovation except improved data collection.

The systemic approach   The second "orthodox" mode of theorizing draws upon a tradition of political sociology that is explicitly sociological. So-called "systemic" political sociology consists of two schools of system building, which differ in style but otherwise have much in common: structure/function theorizing, and general systems theorizing, which can be associated provisionally with the names of Talcott Parsons and David Easton. Parsons has had a widespread influence in political sociology, particularly in the study of political integration and party systems (Lipset & Rokkan 1967). Easton's influence has been less pervasive, except insofar as his work has dovetailed with that of Deutsch (1966) who has been widely influential. But rare has been the textbook in orthodox political sociology that has not begun with a chapter outlining the elements of the political system and its functions according to the frameworks of Parsons and/or Easton and/or Deutsch (e.g., Almond & Coleman 1960, Lipset & Rokkan 1967).

Parsons's (1937) point of departure is the assertion that a conception of individual actors pursuing random ends cannot explain social order. Rather, the social internalization of norms accounts for the orderly outcome of individual choices. In other words, order results from voluntaristic action in which people *accept* normative values to guide their "individual" rational choices. The political system is assigned a relatively limited function within the normative social order. Its role is to organize people to act together in means–end relationships: identifying shared objectives, planning a route to them, and organizing "actors" so that they follow the best route.

Lipset and Rokkan (1967) base their model of cleavage structures, party systems, and voter alignments on Parsons's "theory of action." Their "inventory of potential cleavage bases" follows the schema for classifying the functions of a social system that Parsons devised with Bales and Shils (Parsons *et al.* 1953). Four "functional subsystems" of every society and six lines of interchange between each pair are identified. The tasks of the political sociologist are seen as examining the interchanges between the *I* (Integrative – associations) and *G* (Goal Attainment – the polity) subsystems, the development of interchanges between the *L* (Pattern Maintenance – households, schools) and *I* subsystems, and regularities in the relationships between subsystems *G* and *L* (Lipset & Rokkan 1967:7–8). Parsons's framework therefore provides a research program and a systemic perspective from which to pursue empirical work in political sociology.

The work of Easton and Deutsch has its origins in a different literature from that of Parsons. One of its origins is in the views of biologists who opposed reductionism – that is to say, an analysis of organisms merely as a sum of parts – but who also rejected vitalism. Parsons claims that his system is empirical, but the general systems theory developed by biologists and others makes claims that it is axiomatic (Phillips 1976). But general systems theory as espoused by Easton shares with the Parsonian system voluntarist, holist, and positivist presuppositions.

Easton's *The political system* (1953) rests on the often quoted definition that political science is "the study of the authoritative allocation of values for a society" (p. 129). The "allocation of values" means "all those kinds of activities involved in the formulation and execution of social policy." The political system "consists of a web of decisions and actions that allocates values." "Authority" is that which makes a policy authoritative. This happens "when the people to whom it [a policy] is intended to apply or who are affected by it consider that they must or ought to obey it" (p. 132). A "society" here is equivalent to Oakeshott's phrase "the general arrangements of a society." The political system is not a concrete subsystem, physically and spatially distinct. Rather it is an "abstracted system": politics is not only something done by some people all the time; it is also done by many people some of the time.

Easton sets out his model of political system in the form of a flow diagram. Such diagrams are very popular with political sociologists. The political system embraces all those activities in a society that are concerned with the authoritative allocation of values, as defined above. Some features of the system are concrete; individual actors *make* demands and *choose* to comply with the decisions of authority, for example. But for the most part, the system is abstracted and holistic. Like other "living systems," the political system is an open system in a steady state. The steady state depends on a balance of inputs and outputs. The inputs are supports and demands: support provides the "energy" to process demands and produce outputs in the form of authoritative policies; demands provide the "information" upon which policies are based. The political system is subject to "stresses," which may be due primarily to lack of supports, or to excess of demands, or to outputs which produce negative effects in the environment and "feed back" into the system as more demands and fewer supports. Demands can be controlled in two ways: structurally, through parties and interest groups that "channel" demands in ways functional to the system; and *culturally*, through modifying beliefs, values, and so forth in such ways as to discourage the demands. Culture has importance in terms of its *function* for the system, therefore, rather than the meanings it provides for people.

At first sight, Easton's system seems distinctly structuralist when compared to that of Parsons. Certainly, the metaphysics of "man as subsystem" are not of first-rate importance to Easton! Politics is a matter of process and events rather than of action. But the actor is there as a subsystem in a hierarchy of levels. Ultimately people must accept the values allocated by the political system as well as the right of this system to do the allocating.

The division between behavioral and systemic political sociology, like the more general division between "empirical" and "theoretical" sociology, has given rise to attempts to "link" the two together by means of mediating concepts. That this involves integrating the presuppositionally incompatible has not received much recognition. Two concepts have been particularly attractive: *political culture* and *power structure* (Almond & Verba 1963, Domhoff 1967).

Political culture is a term coined to connect the vocabulary of behavioral with that of systemic political sociology. To Almond and Verba (1963: 13), for example, the political culture of a society is "the political system as internalized in the cognitions, feelings, and evaluations of its population. People are inducted into it just as they are socialized into nonpolitical roles and social systems." The classification system they adopt is drawn from Parsons and Shils (1951). The language used is a hybrid of behavioral vocabulary and the jargon of structural–functional sociology. The major objective is to link together "micro-" and "macropolitics." To Almond and Verba (p. 32), "the relationship between the attitudes and motivations of the discrete individuals who make up political systems and the character and performance of political systems may be discovered systematically through the concept of political culture."

Proponents of power structure research likewise focus on the integration of micro- and macropolitics. But unlike the emphasis on *compliance* in the concept of political culture, power structure implies the dominance of an elite, ruling group, or establishment over political institutions and political life in general. This approach has a longer history than that of political culture. It can be traced back to Mosca, Pareto, Michels, and Sorel at the turn of the century (Runciman 1969 : 86). Its main features lie in rejection of priority to normative values in explaining social order, and stress on the coercion and manipulation of populations by a "power structure." The term itself represents an attempt to bridge the gap between a behavioral and a systemic account – power implying agency, and structure implying, in this usage, a holistic conception of society.

The Marxist–voluntarist approach   There is another "style" of political sociology, though much more internally coherent, which has emphasized just as strongly the nature of politics in "capitalist" societies as a process of domination. This is what I am calling "antithetical" political sociology. It derives from an explicit rejection of the idea, characteristic of "orthodox" political sociology, that the *state* is, in Hegel's sense, a "higher universal" in which the conflicts of *civil society* can be overcome. To the contrary, the state depends precisely *upon* these conflicts, especially that between the two classes – bourgeoisie and proletariat – which define the basic economic interests of capitalist society. According to this viewpoint, the predominant question concerns the situation, interests, and struggles of the social classes engaged in class conflict (see, e.g., Miliband 1969, Moore 1966, Allum 1974).

Until the 1960s this mode of theorizing was not explicitly structuralist, even if it was clearly holistic and positivist. Indeed, the reference to "class struggle" implies an element of voluntarism. And an important strain in Marxist political sociology emphasizes this even now (e.g., Miliband 1977). But a significant attempt to base a Marxist political sociology in an internally coherent theory of politics rather than as a *derivative* of a theory of economy has been overtly structuralist.

The Marxist–structuralist approach This is seen best in the writings of Godelier (1972) and Poulantzas (1973). Godelier argues that two principal contradictions haunt capitalist society: the first being that between bourgeoisie and proletariat, the second – the "basic contradiction" – being that between the growth and socialization of productive forces and private ownership of the means of production. The breakdown of capitalism under this basic contradition is *structural*. It is *not* the result of "class struggle." Human agency has nothing to do with it.

Similarly, Poulantzas criticizes Miliband and other Marxist "voluntarists" by insisting that the state and social classes are "objective structures." Individuals exist only as the "bearers" of "objective instances." The object of inquiry – politics in "capitalist social formations" – is constituted by the "mode of production," an ensemble of "levels" (economic, political, ideological, and theoretical) forming a complex whole, determined, "in the last instance," by the economic level, but in which the economic level *may* not have the dominant role (Poulantzas 1973:14–15). One mode of production is distinguished from another by the particular *articulation* of the various levels. "Structural causality," then, is not a rigorous determination of a *specific* effect, but the production of certain outcomes from an articulation of levels.

An important innovation by Poulantzas has been the rejection of a simple two-class model of cleavage because of its empirical irrelevance. In its place he, and others, have proposed distinguishing "autonomous class fractions" and strata created by "secondary effects of the combinations of modes of production" (Poulantzas 1973:77–93). The danger for Marxist structuralism in these definitional references to political effects created by the fractionalizing of the classic classes is that there is no limit to the proliferation of "fractions" and "strata." There is a return, full circle, to the arbitrary classifications of behavioral political sociology.

By way of conclusion for this section, there are therefore four major modes of theorizing in contemporary political sociology – the behavioral, the systemic, the Marxist–voluntarist, and the Marxist–structuralist – and two attempts to join the behavioral to the systemic – political culture and power structure research. But as illustrated by these last two "types," there has been dissatisfaction in some quarters with the inability of single frameworks to encompass the full range of questions of interest to political sociologists. A number of "areas of challenge" can be identified. An alternative to existing modes of theorizing would have to meet these challenges in order to qualify as an improvement over them.

Areas of challenge

Most theorizing in social science is not reflexive. It does not consider its philosophical presuppositions and their possible limitations in a self-conscious manner (Szymanski & Agnew 1981). Increasingly, however, the sets of presuppositions upon which dominant modes of theorizing in political sociology rest have been called into question in "general" sociology (Giddens 1979, 1981; Lukes 1977; Burman 1979; Sztompka 1982). Moreover, within the field of political sociology itself there have been attempts to "merge" perspectives by means of mediating concepts or subsuming one level of analysis under another (e.g., Effrat 1972, Hechter 1983).

There are three respects in which *all* major modes of theorizing or derivatives of them are radically insufficient for the purposes of political sociology. The first of these is the inability of existing approaches to bridge the gap between the "macroscale" and the "microscale" (the holist and the individualist) without subsuming one under the other. Yet there is an increased upsurge in what can be called "microsociology" that challenges both holist and individualist presuppositions (Knorr-Cetina & Cicourel 1981). This challenge is important because it offers the prospect of grounding analysis of political behavior and institutions in research on interaction in microsocial situations.

The second area of challenge concerns the sufficiency of positivism or intuitionism as "models of knowing." The major models of knowing are predominantly positivist. In the classic polarity, intuitionism is the only alternative available. But this is not the case. The recent revival of interest in scientific realism suggests another model of knowing that has not been given careful consideration in political sociology.

A third area of challenge involves the way dominant modes of theorizing characterize the concept of power. The presupposition of voluntarism is that agents (individuals or groups) are completely self-determining. The presupposition of structuralism is that they are completely determined. In one case power is unlimited, in the other it is nonexistent (Lukes 1977). Lukes (1975, 1977) has pointed to the problems with this dichotomy and how the concept of power can be reformulated so as to avoid the presuppositions of voluntarism and structuralism.

FROM HOLISM AND INDIVIDUALISM TO SITUATIONALISM

In individualistic models of society such as those in behavioral political sociology the concepts used must be understood in terms of the interests, activities, and so on of individual human beings. To the contrary, in holistic models "social facts" are viewed as independent of individuals. The methodological position of recent microsociology challenges both of these stances. Its most radical version claims that social facts are unknown and unknowable unless they can be grounded in knowledge derived from analysis of microsocial situations. But, at the same time interaction in social situations rather than the isolated "individual" is the preferred focus of analysis. Goffman (1972:63) points to this shift in emphasis in an article appropriately entitled "The neglected situation," when he argues that much social research has implied that "social situations do not have properties and a structure of their own, but merely mark, as it were, the geometric inter-action of actors making talk and actors bearing particular social attributes."

A clear statement of the interactional basis to social action was made by Simmel (1971:23) when he wrote that:

> Society exists where a number of individuals enter into interaction. This interaction always arises on the basis of certain drives or for the sake of certain purposes. Erotic, religious, or merely associative impulses; and purposes of defense, attack, play, gain, aid, or instruction – these and countless others cause a man to live with other men, to act for them, with them, against them, and thus to correlate his condition with theirs. In brief, he influences and is influenced by them.

The reality of social situations cannot be predicted, therefore, from know-ledge of the attributes of single actors as behavioral approaches contend. Rather the outcome of social action is tied to specific occasions and to other participants in the situation. Behavior is therefore contingent upon others.

This approach has of course been developed extensively by Schutz and his followers. They identify the "environment," the "context," and the "setting for social action" as the defining settings for social situations. But these are not viewed as "external" to individuals but are themselves constructed in social action (Schutz 1967). From the microsociological viewpoint this is a necessary condition. The primacy given to agents' practical reasoning implies that the most enduring situations and repetitions are those around particular settings and objects (Collins 1981:995).

A critical question concerns the possibility of grounding "macro-order" in terms of "microepisodes," such as those identified by Goffman, Schutz, and others. Collins (1981) has proposed that macrophenomena are made up of *aggregations* and *repetitions* of many microepisodes. Microepisodes are the situated social encounters partly structured by past definitions and yet "always open" to reconstruction (Brittan 1973:84). The notion of social class as a social fact or holistic phenomenon can be rethought in situational terms. If class relations are viewed as consisting of many and repeated situations in which those who own and control the means of production confront those who do not, a social class can be defined by the net interactional experience of distinct sets of individuals in these situations. So macrophenomena can be logically derived from microsituations (Collins 1981:994–6).

But this perspective implies that the whole social process is somehow up front, and derivable from the intended action of competent and knowing actors who face no *macro*level constraints. Giddens (1979), and others, however, argue that the long-term formation and potential transformation of institutions should be viewed in terms of the *unintended* consequences of social action which always limit the capability and knowledgeability of actors. The aggregation approach of Collins therefore probably needs elaboration in terms of unintended consequences that condition whether in fact the parts (situations) do constitute the whole (macro-order).

Knorr-Cetina (1981) proposes a third way of integrating microepisodes into macro-order. She argues that "unintended consequences" may be a redundant concept "if the interrelation of scenes of action by and for agents construed through representations of mutual knowledge, intentions, projects, interests, etc., are given adequate consideration" (p. 33). Following this strategy, the macro is conceived of as actively construed and constructed *within* microepisodes rather than as an *emergent* phenomenon composed of the unintended effects of microevents. In other words (p. 34):

> the macro appears no longer as a particular *layer* of social reality *on top* of micro-episodes composed of their interrelation (macro-sociologies), their aggregation, or their unforeseen effects. Rather it is seen to reside within these micro-episodes where it results from the *structuring practices* of agents. The outcomes of these practices are representations which thrive upon an alleged correspondence to that which they represent, but which at the same time can be seen as highly situated constructions which involve several levels of interpretation and selection.

It is clear that there is some disagreement about how to move from microepisodes to macro-order. But it is fortunate that at long last considerable attention is being given to the problem at all. In the struggle between holistic and individualistic models of society in the recent past, the micro–macro issue was never seriously addressed. There is also a potential for resolution.

A synthesis of the three positions is in fact possible. The macroscale is palpably represented in the routines and practices of everyday life. But it is itself the aggregate product of the consequences, intended and unintended, of microscale situations. Aggregation, unintended consequences, and representation are all at work rather than just one of them.

One can indeed go somewhat further in "concretizing" the macro-order without abandoning the microsociological perspective. Bhaskar (1979) and Burman (1979), for example, argue that the causal powers of individual persons are alienated to a considerable degree by the existence of practices which presuppose an objectivated macro-order. To Burman (1979: 364), for instance,

> Human beings produce their social reality, and that social reality becomes objectivated and acts back upon the individuals. It is dialectical

in that, on the one hand, the *objectivated social facts* have been humanly constructed, humanly maintained and should, provided they are not reified, be perceived as expressive of subjective human intentions; and, on the other hand, the externalized activity is to some extent shaped by and infused with objectivated contents.

The capability and knowledgeability of specific persons are therefore always limited by the structures (rule/resource sets) of externalized activity. The practices of others, both known *and* unknown, living *and* dead, near *and* far, provide the stuff of *structures* of doing, knowing, and feeling that in *concrete* institutional and cultural forms help define the microsociological settings in which specific persons think and act (see Ch. 3 for more discussion).

FROM POSITIVISM AND INTUITIONISM TO REALISM

For many years positivism and intuitionism have been presented as the competing models of knowing in social science. But a third position, that of scientific realism, has recently received considerable attention from those dissatisfied with positivism but finding little appeal in extreme intuitionism (Bhaskar 1979, Keat 1971, Keat & Urry 1975). The major issue with positivism for realists is whether the real world is exhausted by what is given empirically. The major problem with intuitionism for realists, and of course for others too, lies in its acceptance of discursive reasons as the sole basis for its "action" descriptions.

Realists have a *naturalistic* conception of people as rule-creating and rule-following agents, in contradistinction to other models. Modern, as opposed to medieval, realism can be defined as the conviction that scientific inquiry is capable of revealing "the truth about the world" and that "this truth cannot be assigned limits corresponding to the limitations of the experimental procedures employed in its determination" (MacKenzie 1977:37). The central concern of realists is thus neither with procedures and "law seeking" nor with the self-reported reasons of individual actors. Rather it lies in discovering the mechanisms (or causes) that produce human behavior. These may well include reasons, as the emphasis on "rule following" suggests, but practical reasons rather than discursive or theoretical ones. This is an important difference between realism and intuitionism: realists seek to explain by reference to natural necessity and things defined in time and space; intuitionists seek to explain *entirely* by reference to the ideas and psychological processes of individuals and groups *free* of time and space (Harré 1970).

The consequences of adopting realism as a model of knowing are several, and congruent with a shift to "situationalism." First, the realist conception of cause leads towards a concern with *contingency*. For realism, causality involves not a relationship between discrete events ("cause and effect") but the *causal powers* and *liabilities* of objects and people (Harré & Madden 1975).

A causal claim is, on this view, what an object or person is like and what it *can* do. Causal powers and liabilities can be attributed independently of any specific pattern of events. Whether they are *activated* depends on conditions whose presence and form are contingent. As Sayer (1984: 99) puts it: "although causal powers exist necessarily by virtue of the nature of the objects which possess them, it is contingent whether they are ever activated or exercised." This means that, depending on the conditions, the operation of the same causal power (which for people includes ideas, beliefs, and reasons) can produce different outcomes. Explanation thus entails more than identifying regular association after the fashion of much statistical analysis, or giving causal priority to theoretical entities after the fashion of structuralism (Vuillemin 1984).

Secondly, realism leads to a different empirical research strategy from that of positivist social science. Rather than conducting *extensive* research, realism encourages *intensive* research (Harré 1979: 132). In intensive research the major questions concern how some causal relations work out in a particular setting. Extensive research is concerned with finding some of the typical attributes and common properties of a large, spatially dispersed population. Important methods of extensive research are questionnaire surveys and census tabulations followed by statistical analysis. Intensive research methods are mainly qualitative, participant observation and informal interviews, or a combination of local quantitative and qualitative analysis.

Perhaps the major difference between the two types of research design is the focus of extensive research on what Harré (1981 : 147) terms "taxonomic collectives": groups whose members are not *substantially* connected in any way but who belong to population categories provided by the researcher or a census organization. In other words, as Sayer (1984: 92) states it, "generalizations are indifferent to structures. Even where they refer to like-constituted entities they say nothing about whether each individual is independent of or connected to any other."

Thirdly, and finally, realism leads to what can be called a "geohistorical" rather than a "psychosocial" conception of social science. This involves accepting the importance of a nonfixed and hence nondetermining geographical environment for social action. The *presence* and *absence* of the contingently related conditions under which causal mechanisms are activated depends on the geographical incidence and arrangement of those conditions over time. Both positivism and intuitionism direct attention away from such considerations towards either investigation of presumptively "nationwide" or universal psychological and social processes that are independent of time–space constitution *or* solipsism. The influence of Max Weber, or Weber's "heritage" (Zaret 1980), on both camps is apparent. As Stoianovich (1976: 143) argues in his examination of trends in recent French social science "Weber's ideas have given succor to people seeking a way out of geohistory." What is so obvious about Runciman's (1969) discussion of positivism and intuitionism in social science is the extent to which the terms of discourse for both are those laid down by Weber. And in each case a "psychosocial" rather than a "geohistorical" view of social science prevails (for a recent good example, see Saunders 1985).

REFORMULATING THE CONCEPT OF POWER
A third and complementary "area of challenge" is the way in which power is defined in dominant modes of theorizing. This has two aspects to it. One is the corollary to the earlier discussion of "causal powers." If one can attribute causality to agents, then one is also attributing power to them. Power in this usage then presupposes human agency. But it is important to recall that whether or not causal powers are activated depends on *conditions* whose presence and form are contingent (Walker 1985). Thus, rather than the power *versus* structure terms in which most social science is conducted, a "dialectic" of power and structure is at work. Ferrero (1981:41), for example, sees the state *and* its subjects bound together in a relationship similar to that of the tragic biblical figures Cain and Abel. In the conventional view Cain represents the men destined to command (the state), Abel represents the men destined to obey (the populace). But to Ferrero the relationship is more complex: "If the subjects always have fear of the Power to which they are subjugated, the Power always has fear of those at its command. Cain had fear of Abel and for this finished by killing him" [my translation]. The relationship between state and society, therefore, is not a one-way street. Indeed, from this point of view, Abel can challenge, if not slay, Cain! But under what conditions can this happen?

Conditions present themselves in several ways. Some operate through an agent's reasons, others limit the agent's ability or opportunity to act (Lukes 1977:13). Together these present what Lukes (p. 20) calls "structured possibilities which specify the powers of agents." As a consequence Lukes is led to reject behavioral and, more generally, positivist accounts of power – the former because it "focuses exclusively on the narrow thread of actualized possibility, rejecting the unactualized as of no explanatory significance," and the latter because "at least in its narrower forms ... it systematically devalues the explanatory role of counterfactuals [what could have occurred] and the value of evidence needed to support them" (p. 29).

A second aspect of the challenge to the definition of power in dominant modes of theorizing follows from this. It involves the identification of power solely with the state and its political system or with individuals as isolates. Foucault (1975) for one has argued that power emerges from "local arenas of action" rather than the "state apparatus" or political system. To Foucault, power should be viewed as a "microprocess" of social life or pervasive feature of concrete, local transactions (Foucault 1975:29–33). Many microsociological studies reinforce this viewpoint. Families, prisons, asylums, law enforcement agencies are all scenes of power struggles no less complex than those of the large political system (Laing 1956, Cicourel 1968).

In two respects, then, conventional views of power are deficient. The first is the focus on either power *or* structure with power referring to overt behavior rather than contingent outcomes. The second is the arrogation of power to the state or political system. An alternative view sees power as productive and enabling as well as repressive, and a feature of all social interactions rather than just those characteristic of the state or political system.

The contemporary condition of political sociology

The term "crisis" may suffer from overuse in recent social science. But I hope that in the preceding pages a general outline of the real crisis facing political sociology has come into focus. The major purpose has been to characterize the dominant modes of theorizing in terms of their fundamental presuppositions, and point to areas where these presuppositions are now under challenge. This is in its essentials an intellectual crisis. But it is in a sense also a political crisis. Both orthodox and antithetical political sociologists have hitched their intellectual wagons to the national state, not only as the focus *of* political activities but also as the focus for *explaining* those activities. But what the challenges suggest is that the way political activities are explained should be radically changed. In particular, approaches such as those characteristic of political sociology should be replaced by ones in which a central focus is on the microsociological.

Over recent years a dramatic reconstruction has *not* been visible within the field. Rather, two main trends have been apparent. One has been a shift away from the voluntarist–holist–positivist mode of theorizing, and a correlative expansion of behavioral political sociology, particularly the "economic theory of democracy" (Budge *et al.* 1976, Poole & Rosenthal 1984). The second has been the relative expansion of the structuralist–holist–positivist mode of theorizing (Wallerstein 1974, Paige 1975, Skocpol 1979). These two trends can be understood as intellectual reactions to the failure of the well-established grand theories (Marxism–Leninism, Parsons) to provide a meaningful and useful context for empirical work. "Going to extremes" might seem to be a satisfactory alternative (e.g., Hechter 1983, Wallerstein 1983). They are, of course, also political reactions. The normative division of political sociology into "orthodox" and "antithetical" parts is now reformulated in a much clearer demarcation of modes of theorizing. They are no longer "fuzzy sets."

But neither of these trends deals with the fundamental challenges to contemporary political sociology *in toto* raised by recent work in "general" sociology and philosophy. The next chapter proposes a reformulation of political sociology that is explicitly sensitive to the problems with conventional political sociology and the challenges they pose.

3 A theory of place and politics

The theories of social organization and social change that have been dominant in Western, particularly American-influenced, social science since the Second World War have had no or only a limited role for a concept of place. Why this has been so is the subject matter of Chapter 5. There have been certain genres of social science, however, to which a concept of place has been central. One thinks, in particular, of French social geography; French electoral sociology; the *Annales* school of social history; American regional sociology, in the fashion of the "North Carolina school"; American political sociology in the "group-ecological" tradition of Key; and, perhaps above all, historical and cultural geography, and American cultural anthropology. By and large these strands of place-based social science have remained intellectually detached from one another and marginal to mainstream social science. Especially in the English-speaking world, but also more generally because of that world's global cultural dominance, these genres have been marginalized or crushed by the dominant mode of analysis that still prevails today – "universalizing, empiricist, sectioning off politics from economics from culture, profoundly ethnocentric, arrogant, and oppressive" (Wallerstein 1978 : 5). Particularly in the 1950s and 1960s, at the height of *Pax Americana*, they almost disappeared entirely; American imperialism and a concept of place in American social science have been mutually exclusive (Agnew 1983). Some genres have also suffered in the past from conservative political connotations (Ch. 5 should make clear why), and the intellectual stigma in some quarters of historical–geographical specificity as opposed to the scientific kudos of large-scale, cross-sectional and "law-finding" social science.

In this chapter an attempt is made to draw on recent work, particularly that of Bourdieu, Giddens, and Pred, to provide a theoretical framework for political sociology in which place is central. Such a theoretical framework is largely missing or at the most implicit in the place-based genres of social science mentioned previously. The framework builds on the critique of both "orthodox" and "antithetical" political sociology provided in the previous chapter. As outlined there, a major dilemma for contemporary social science is its inability to manage what might be called the "microsociological challenge" to macrosociology and individualism (Knorr-Cetina 1981) without dramatically reconstructing social theory. One way out of the dilemma is to build an approach focused upon the concept of place as the nexus of the structuring of social relations.

The chapter proceeds as follows: firstly, the problem of defining place is discussed; secondly, a place-centered approach to social science is described; and thirdly, the relationship between place and the state is examined.

What is place?

Place is one of those "contestable" concepts (Gallie 1955-6) whose application is a matter of dispute. In geography, it is often used synonymously with location, point, area, or space. This confusion has led Giddens (1983:79) to suggest that the term "locale" substitute for place in the sense of indicating the physical settings in which social relations are constituted. The concept of place as context for social relations has suffered particularly from its assimilation in sociological discourse to the concept of community (see Ch. 5). But an emphasis on places as the physical contexts for action has long been characteristic of microsociology and much humanistic geography (Ley & Samuels 1978). This perspective represents the position that in order to *explain* human behavior one must deal with the "microepisodes" of everyday life and their embeddedness in concrete milieux or contexts. To Giddens (1983:79), therefore, "Locales are not just points in space in which action occurs, any more than time is a series of intervals into which action is somehow inserted." There is nothing particularly novel about this observation. Place has long had the sense of setting for interaction that Giddens now wants to associate solely with the term locale (e.g., George 1966, Tuan 1974, Raffestin 1981). Only in recent intellectual history has the idea of place *solely* as location become predominant. Moreover, in substituting "locale" for "place" there is a danger of missing out the aspect of place captured by location. This involves not only everyday social practices but the long-run siting of locales through the distribution of resources and the physical construction of settings. In Chapter 5 it is argued that this is no accident. The devaluation of place has been semantic as well as conceptual as certain features of social science have led to the removal of geography from an *intrinsic* role within social theory.

An emphasis on the contextuality of action, a notion central to the definition of place adopted here, has long been characteristic of microsociology. I have in mind approaches such as symbolic interactionism, cognitive sociology, ethnomethodology, social phenomenology, ethogenics and ethnoscience. This literature stresses the fact that people do not experience life in the abstract context of "mass society." Their knowledge is acquired, and they live their lives, in the context of "social worlds" dominated by the perspectives of different "reference groups," in which meaning is attributed to acts and events through communication and interaction with limited numbers of people (Shibutani 1955, S. J. Smith 1984). In everyday life such social worlds provide the boundaries for social learning and interpretation (Lefebvre 1971, Rochefort 1983, Tabboni 1985). This is as true for "cosmopolitans," people with an orientation towards a wider world, as it is for the mass of people, "locals," people whose interests and definition of life are locally oriented (Merton 1968). Even for "jet-setters," reference groups and locale-specific "significant others" define the rhythm of their movements from Acapulco to Aspen to Gstaad to Cannes. When in a locale, so to speak, they follow the routines and rituals of that locale. Of course, most cosmopolitans have rather more limited geographical itineraries and are tied to a few dominant reference groups and one set of locales at a time. Locals are even more socially, and spatially, constrained (Dahmann 1982).

But as Giddens and others have pointed out, microsociology often misses the impact on the constitution of action of the *longue durée* of structured social practices. There is a sense in which "locales" could be anywhere. However, they are not. They are *located* according to the demands of a spatially extensive division of labor and global system of material production and distribution. This is the macro–order of objectivated social facts referred to in Chapter 2. The "face-to-face society" of the locale in which action is embedded is in its turn embedded within a wider "territorial society" (Laslett 1956). This is important because, as Pred (1984: 283) argues:

> In industrialized countries the spatial and social division of labor occurs at a macro-level within a system of places while retaining a local component. Thus, especially in capitalist countries, but also in command-economy countries, the production and distribution projects occurring within a local area are directly or indirectly connected to the dialectics of more macro–level structuration processes.

Therefore, and this point is played down by Giddens, place is not just locale, as setting for activity and social interaction, but also location. The reproduction and transformation of social relations must take *place* somewhere. Pred puts it as follows (p. 279):

> Place ... always involve an appropriation and transformation of space and nature that is inseparable from the reproduction and transformation of society in time and space. As such, place is not only what is fleetingly observed on the landscape, a locale, or setting for activity and social interaction (Giddens 1979: 206–7; Giddens 1981: 39, 45). It also is what takes place ceaselessly, what contributes to history in a specific context through the creation and utilization of a physical setting.

But place is also more than an "object" (Faccioli 1984). Concrete, everyday practices give rise to a "structure of feeling," to use Williams's (1977) phrase, or "felt sense of the quality of life at a particular place and time" (Pred 1983: 58). This sense of place reinforces the social–spatial definition of place from *inside*, so to speak. The identification with place that *can* follow contributes yet another aspect to the meaning of place: one place or "territory" in its differentiation from other places can become an "object" of identity for a "subject." This is *not* the same as community in the sense of a way of life based on a high degree of personal intimacy and sociability. But of course this could be present also.

Interwoven in the concept of place adopted here, therefore, are three major elements: *locale*, the settings in which social relations are constituted (these can be informal or institutional); *location*, the geographical area encompassing the settings for social interaction as defined by social and economic processes operating at a wider scale; and *sense of place*, the local "structure of feeling." Or, by way of example, home, work, school, church, and so on form nodes around which human activities circulate and which *in toto* can create a sense of place, both geographically and socially. Place, therefore, refers to discrete if 'elastic' areas in which settings for the constitution of social relations are located and with which people can identify. The "paths" and "projects" of everyday life, to use the language of time-geography, provide the practical "glue" for place in these three senses (Pred 1984). To the extent that places are similar in these respects, interconnected and contiguous one can refer to a "region" of places (Cox 1969b). In that situation the sense of place can be *projected* onto the region or a "nation" and give rise to regionalism or nationalism. The sense of place need not be restricted to the scale of the locality.

The question of how to define place has exercised geographers and others for many years (Tuan 1974, Raffestin 1981, Scivoletto 1983, Strassoldo 1983, Muscarà 1983). In their approaches to it, one or other of the three elements has tended to predominate. For example, economic geographers have tended to emphasize location, cultural geographers have been centrally concerned with sense of place, and a few humanistic geographers have concerned themselves with locale. Rarely have the three aspects been brought together. The central focus here on locale, because of its sociological importance, leads to an emphasis on place as synonymous with *locality*. But the incorporation of location into the definition of place implicates processes at other scales. A key tenet is that the local social worlds of place (locale) *cannot* be understood apart from the *objective* macro-order of location and the *subjective* territorial identity of sense of place. They are all related; if ultimately locale is the most central element sociologically it must be grounded geographically. In other words, locale is the core geosociological element in place, but it is structured by the pressures of location and gives rise to its own sense of place that may in certain circumstances extend beyond the locality.

Place as process

In recent years there has been a convergence between the work of some human geographers and that of some social theorists. This is marked by a common rejection of the duality between determinism and voluntarism that lies at the heart of most modern social theories. In harking back to the critique of Gurvitch (1955, 1963), they all ask: can this duality be overcome by dialectically recombining social structure and human agency? A theory of "structuration," in primitive form first proposed by Berger and Luck-mann (1966), but more recently found in the more sophisticated models proposed by Giddens (1979, 1981), Bourdieu (1977), and Bhaskar (1979), has been espoused as a solution. The human geographers Pred (1983, 1984) and Thrift (1983) have taken up this approach and, in combination with the time-geography of Hägerstrand (1970), have put forward a structurationist perspective on place. In this section the "theory of structuration" is reviewed and then, following Pred's (1983, 1984) approach, developed as a place-centered social theory.

THE STRUCTURATION PERSPECTIVE

Thrift (1983) identifies the core of the dilemma that has inspired the growth of the structurationist perspective by quoting from Sartre (1960, translation by Ferrarotti 1981 : 23):

> That Valéry is a petit-bourgeois intellectual is beyond doubt. But all petit-bourgeois intellectuals are not Valéry. The heuristic inadequacy of Marxism – and, let us add, of traditional biographical method – is contained in these two statements. To grasp the processes which produce the person and his production, within a given class and society at a given historical moment, Marxism lacks – and so does sociology – a hierarchy of mediations ... [one must] find the mediations which can give birth to the concrete, singular life and the real historical struggle, out of the general contradictions of the productive forces and the relations of production.

The various structurationist theorists have contemplated this problem from different angles and approached its resolution in different ways. But they share a set of common concerns and similar proposals.

The first feature of structurational sociology is its antifunctionalism. All of its advocates, most especially Giddens, regard functionalist "explanation" as a misnomer. Much of contemporary social science, they argue, is char-acterized by a set of interrelated errors such as attributing "needs" to societies, imputing a teleology to societies, and the assumption that societies are functionally ordered and integrated. The other features of "structur-ationism" stem from the critique of functionalism.

The second feature is a commitment to a synthesis of structural (objecti-vist) and voluntarist (subjectivist) perspectives within one overarching and internally consistent framework. Structural approaches are criticized for treating human practices as derivative of social structure. Voluntarist approaches are regarded as likewise problematic because as Bourdieu (1977 : 81) puts it:

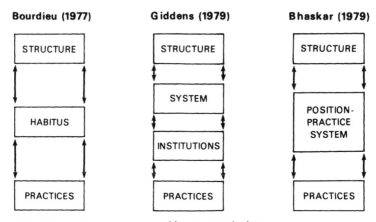

Figure 3.1   The mediating concepts used by structurationists.

interpersonal relations are never, except in appearance, individual-to-individual relationships ... the truth of interaction is never entirely contained in the interaction. This is what social psychology and interactionism or ethnomethodology forget when reducing the objective structure of the relationship between the individuals to the conjunctural structure of their interaction in a particular situation or group.

Rather, social structures are both constituted by human practice and, concurrently, the medium for this constitution. Life is essentially *recursive* (Giddens 1979). By means of *lifelong* socialization, and through the limits set by the physical environment, people *draw* upon social structure. But while they do this they are also reconstituting that structure. Mostly it is an unconscious reproduction. But people are to a certain extent "capable and knowing agents"; they are not "cultural dopes." Therefore, they have the possibility of transforming social structure at the same time that they are products of it.

Authors differ in the mediating concepts they use to relate structure and agency (Fig. 3.1). Bourdieu (1977) inserts a "dialectical" level between social structure and human practices. This is called the "habitus" of cognitive, motivating and "reason-giving" structures that confer dispositions on people. The habitus mediates between objective life-chances on the one hand, and strategies used in interactions on the other. By way of example, each class has a particular habitus that results from common material conditions and generates a common set of expectations.

Giddens (1979) sticks with more familiar sociological terminology, but gives the terms he uses very specific meanings. Thus structure is limited in its meaning to "rules and resources." A concept of "system" is added as "reproduced and regular social practices." "Institutions" are the specific organizational settings in which systems operate.

Bhaskar (1979) offers the concept of a "position-practice system" as his solution to the problem of mediation. He argues (p. 51) that:

> It is clear that the mediating system we need is that of *positions* (places, functions, rules, tasks, duties, rights, etc.) occupied (filled, assumed, enacted, etc.) by individuals, and of the *practices* (activities, etc.) in which in virtue of their occupancy of these positions (and vice versa) they engage.

The third major feature of the structurationist perspective is its focus on practical reason rather than discursive reasons in accounting for human action. Practical consciousness is "tacit knowledge that is skillfully applied in the enactment of courses of conduct, but which the actor is not able to formulate discursively" (Giddens 1979:57). This is not to rule out "explanation by reasons," but to argue that the reasons an actor gives for an action are not *necessarily* the real ones (Fay 1978). More to the point, however, the focus on practical reason reflects an emphasis on the practical grounding of human action *and* social structure.

The fourth, and final, feature of the structurationist position, one that follows from the focus on practical reason, is that time and space are *central* to the construction of social interaction and, consequently, to social theory. To Giddens (1979:54), for example, "social theory must acknowledge, as it has not done previously, time–space intersections as essentially involved in all social existence." Practice is necessarily situated in time and space. This is one link to social structure. Places in which activities occur are the product of institutions which are in turn produced by structure. Place, then, provides the context in which agency interpellates social structure. Rather than epiphenomenal to society, place is central to its structuration.

STRUCTURATION AND PLACE

To date, however, theorists of structuration have not followed through on the *geographical* implications of their perspective. As Pred (1983:46) has noted:

> nobody identifiable with the structuration perspective really has succeeded in conceptualizing the means by which the everyday shaping and reproduction of self and society, of individual and institution, come to be expressed as *specific* structure-influenced and structure-influencing *practices occurring at determinate locations in time and space.*

He goes on to propose that it is possible to integrate into the theory of structuration the concepts of Hägerstrand's (1970) time–geography to capture "the material continuity and unbroken time–space flow of the structuration process." The two basic building blocks of time–geographic language are "path" and "project." The path concept refers to the actions and events that *consecutively* make up the existence of individuals. A project consists of a *series* of activities needed to complete an intended behavior. Each activity requires the convergence in time and space of the paths of involved individuals. Social reproduction and individual socialization, the processes of structuration, take place through the intersection of individual paths with institutional projects occurring at specific sites (workplace, church, school, neighborhood, etc.).

Within any place it is the *dominant* institutional projects, the ones which draw most on available time resources, that have the most impact on socialization and reproduction. In most places the dominant projects involve local material production and distribution. However, the projects of religious, political, and other institutions can often be dominant, or at least important, in structuring social relations (Pred 1984:283).

An alternative way of thinking about place as "historically contingent process," as Pred (1984) puts it, is to view social activity as a *discourse* in which the inhabitants of a place are active participants. Williams (1977) has done the most work in providing the basis for this interpretation of structuration in place. He argues that human practices give rise to a set of expectations and meanings that in turn guide practice. This discourse, or hegemony, changes as human practices change. The prevailing power relations at the center of local social structure establish and reproduce the realm of action and everyday practices. Concomitantly and through everyday and routine practices, power relations are themselves reproduced.

One danger of this approach, but one which Williams avoids, is that the *active* nature of discourse is played down. Socialization is a process of resistance as well as domination. Within places, certain locales can be the settings for "counter-institutions" (Thrift 1983:41) that challenge the ruling orthodoxy of a place. Examples would include any institutions that operate through practices that are distinctive from *and* challenging to the local hegemony.

Another danger is that the "mobilization of bias" inherent in a given setting, the favoring of one group over others in the routines of local social life and politics, will not be seen as a limiting condition on the understanding of the existing social order by groups present in that setting. Variability in the availability of knowledge and the biases of language are particularly important barriers to the extension of the understanding necessary to political challenge (Thrift 1983:45–6).

Except for conventional agricultural settlements, stereotypically perhaps the European feudal village, the "time–space distanciation" of social interaction involves the expansion of social relations beyond the local. The most important factor in this process has been the growth in the social division of labor and the emergence of a class society (Giddens 1979). Particularly for the middle class, interaction has become more often "at a distance." Phenomena such as commuting and leisure travel have stimulated this. But more fundamentally, the increasingly global organization of production and the increased "homogenization" of human practices through the influence of mass media, educational systems, and the "surveillance" (Giddens 1981) of national governments over their populations have helped to make the practices reproduced in different places more and more alike.

But these tendencies do not mean that place, as a particular locational and territorial intersection of settings for social action, has lost coherence or meaning. The strongest forms of bonding are still local: a village or town, particular valleys or mountains. In many parts of the world when asked where they come from, or where they were formed or belong, people respond in terms of local reference (Hummon 1986). Common experiences engendered by the forces of "nationalization" and "globalization" are still mediated by local ones. Most people still follow well-worn local paths in their daily existence. Though national or global "issues" have increased in number and significance relative to local ones, they take on *meaning* as they relate to local agendas. Consequently "local" politics is no longer solely about local issues. But all issues, whatever their source or pervasiveness, are meaningful or important only in the context of outlooks derived from everyday life (Agulhon 1983).

Perhaps most significantly, the social division of labor takes a spatially differentiated form. Whether decisions concerning production and distribution are made locally or between places, there is an unevenness in the spatial distribution of investment, skills, input sources and markets (Pred 1984: 284). The social complexity of places may well have increased rather than decreased as a result of this. Urry (1981: 464) argues that "changes in contemporary capitalism are at present heightening the economic, social and political significance of each locality." Different places have different relationships with the international economy: some operate as the "production outposts" of multinational firms; some are more insulated from the international economy, with an orientation towards regional or national markets but increasingly threatened by "foreign" competition; others are themselves bases for "international empires" (Massey 1984: 299). But local areas rarely bear the marks of only "one form of economic structure. They are products of long and varied histories. Different economic activities and forms of social organization have come and gone, established their dominance, lingered on, and later died away" (Massey 1984: 117).

One trend in particular – increased social and geographical mobility – rather than decreasing the significance of place, as often alleged, may indeed have enhanced it. The tendency to sequestration on the part of social groups may follow directly from the decline of status-markers other than residential differentiation (Sopher 1973). Space, it seems, is *increasingly* divided into enclaves in which distinctive social groups carry on parts of their everyday lives in isolation (Suttles 1972; Giddens 1981; Lefebvre 1971, 1976). In particular, the spread and widespread acceptance of private property rights in land, houses, and things "within one's zone," *stimulate* a concern with "defensible space" (Evans 1984). One's property, representing as it does personal autonomy, social esteem, and exchange value, must be protected from destabilizing events and people (Agnew 1981c). Especially, but not only, in the United States, "good" neighborhoods, suburbs, or places in general, are economically viable ones – most people are home owners; property values are increasing; mortgages, bank loans, and insurance coverage are readily available: the area is "stable." "Bad" neighborhoods or places are ones with the opposite characteristics (Cox 1978, Agnew 1984b). By dint of living in a society in which the metric of economic transactions, exchange value, is the dominant measure of worth, people come to accept it and use it as a *natural* measure of place-valuation and place-definition. "The logic of liberal capitalism" rather than undermining place through a decline in what Giddens (1979) terms "presence availability" (local ties), has provided new significance for it as people establish and defend paths and projects against capitalism's "relations of absence" or extra-local ties.

Logan (1978) argues that a major outcome of the process of place creation is a collective effort within places to pursue personal advantages through competition between places. Drawing on the literature in human ecology he proposes (p. 404) that "Places with early advantages, by making full political use of their superior resources, can potentially reinforce their relative position within the system of places." Places can therefore become collective actors. "People and organizations are ... bound together by the places in which they live or have invested" (p. 409). Places are thus "communities of fate" (Stinchcombe 1965), in which prospects for social and economic rewards are tied to the advantages and disadvantages of the places – for example, levels of income, availability of housing and employment, rates of taxes, levels of public services, etc.

This is not to say that all groups within a place have the *same* interests in development, or that all interests are equally represented. Much empirical work suggests quite the opposite. But "spatial differentiation does in practice imply inequalities among places, and thereby advantages and disadvantages to the persons and organizations whose fortunes are linked to specific places" (Logan 1978:414). This explains in part the desperate "beggar your neighbor" competition for new industry and employment among American municipalities (Bluestone & Harrison 1982) and the common use in the United States of place of residence to support the status claims of individuals (Perin 1977, Berry *et al.* 1976). It also explains in part the different appeals of different national political parties in different places. The ideologies and policy proposals of different parties appeal more success-fully in some places than in others as they strike responsive or nonresponsive chords among dominant sectors of local electorates (Trigilia 1981, Bensel 1984, Brusa 1984a). In these circumstances, national parties can be viewed as instruments for place as much as social group or personal advantages, although these are necessarily implicated too.

But it is not just as the location of important locales that place has social significance. It is also as a seat of sentiment or "structures of feeling." The "particular activities" that give rise to a structure of feeling, Pred (1983:57) argues, "are always enacted in a specific locality, or place." Now, many of these do involve what can be called "class" or "generational" experiences and associated sentiments (Bhaskar 1983). But these are grounded in the structuration process of historically specific situations. Even apparently "nonplace" social movements, such as the American civil rights movement and the American women's movement, often arise more strongly in some places than others because the social context directs attention to certain issues and facilitates their expression. For example, the abolitionist movement (to abolish slavery) in 19th-century America was stronger in Western states such as Kansas than elsewhere; the 19th-century American women's movement originated in upstate New York. In this century similar patterns are apparent. For example, the nuclear weapons freeze movement started and still has its greatest strength in New England. Some other examples are discussed in a section of Ch. 4 on new political movements. Pred (1983:56) elaborates in time–geographic terms as follows:

> an individual acquires a structure of feeling by having her path exposed to news of particular political–historical events by word-of-mouth, the printed word, or the modern media; partly by the everyday intersection of her path with time–space specific institutional projects which also require both the path intersections of other persons (some of whom belong to the same generation or class) and common interaction with objects . . .; and partly by the constraints and possibilities imposed on her other forms of project participation, and thereby knowing, by fixed commitments to dominant institutional projects.

Of course, "sense of place" is often viewed as passé. Partly this has happened (and the burden of Ch. 5 is to demonstrate this) because place has been confused with the sociability of the ideal-type community that has passed from the modern world. But partly it reflects the increasingly fragmented character of everyday experience and the increased realization that the institutional projects and rules one encounters every day do not originate locally but in distant "seats of power" (Pred 1983:63). However, rather than engendering a total "placelessness," as writers such as Relph (1976) contend, there are structures of feeling or place identities that continue to characterize locally circumscribed areas both because of resistance to "relations of absence" *and* the physical uniqueness of individuals' paths (Lévy 1984a). Indeed, the "politics of territory" that has become increasingly obvious in contemporary Europe reflects in part an identity with places and their problems that is a self-conscious response to the increased ratio of extralocal ties (Osmond 1977, Orridge & Williams 1982). Although in this case, as in others, "sense of place" is now often at a scale beyond that of the locality. Moreover, the sociability of community defined by place may not have disappeared, but rather changed its form (Eyles 1985). For example, communal festivals and church attendance may be displaced by union meetings, cafes, and political clubs. A sense of community can present itself through different manifestations (Thrift 1983:47).

Place, therefore, is not the *outcome* on a map of an abstract social process beyond place. Place refers to the *process* of social structuration. As Massey (1984:117) puts it: "most people still live their lives locally, their consciousness is formed in a distinct geographical place." A major implication follows: "At any one time different areas may be changing in contrasting ways; different battles are being fought out, different problems faced." Geography still matters:

> No two places are alike. People who only stop over at airports (to and from conferences, perhaps) might get the impression of a pervading similarity, but any deeper exploration will soon dispel that illusion. Geographical variation is profound and persistent, and this is true even within a country as apparently homogeneously developed as the United Kingdom. We talk of the "peripheral regions", but they are all very different. The term "inner cities" includes Merseyside, Glasgow and London, as different as can be in culture, political history and the character of the labour movement. Any train journey across country provides ample evidence: different kinds of people get on and off at Lancaster, Grange-over-Sands and Barrow-in-Furness (Massey 1984:117).

Place and the state

What goes on in places cannot be understood without reference to the "outside forces" that help define those places. On the one hand, a place contains landlords, workers, managers, and peasants whose interests, together with those of other people in those categories elsewhere, are represented at other levels in a political system, producing a changing pattern that to a certain extent determines intergroup relations locally. On the other hand, wars, recessions, coups d'état, and other "larger events" are absorbed into the life of places and given social meaning.

One of the most important "outside forces" has been the modern territorial state, especially important in the context of place and politics. State formation has been a great and powerful process. Expanding geographically from distinctive core areas, states have penetrated, standardized, and incorporated previously independent local systems of power into state-based ones (Tarrow 1977). The process has been international and historically specific. After AD 1500, national states supplanted clans, empires, cities, churches, federations, and other kinds of groupings as the dominant organizations throughout the world. They grew first in Europe, expanded into Europe's dependencies, and eventually took over the rest of the world.

State building in Europe involved the construction of new political units rather than the straightforward transformation of existing ones. Often the rulers of one area with certain advantages in resources or organization extended their control to adjacent areas by conquest, subversion, inheritance, marriage, or alliance (Pounds & Ball 1964, Raffestin 1981). Italy grew up around the territories of Savoy, Germany around the lands of Brandenburg–Prussia, Spain around Castile. These cases are especially telling because in all of them effective unification or integration under a central government came only recently (Tilly 1975). Even in the United States, the prototypical "first new nation" (Lipset 1963), where political democratization accompanied state building, political integration was far from consensual and, indeed, has been marked by bitter intersectional and interethnic conflicts.

State building was part and parcel of the creation of the modern world-economy. This is essentially a system based on massive capital accumulation. From its origins in the 16th century, this system has engulfed the world (Wallerstein 1974). States have been its major agents. The states of Western Europe and North America, in particular, have been dominant within it. As a result of such dominance they have acquired certain unique characteristics: social structures marked by elaborate divisions of labor, the dramatic decline of employment in agriculture, large and finely differentiated middle classes, the appearance of secular–rational political ideologies, and the institutionalization of representative politics and its major instruments, political parties (Burnham 1984 : 113).

Most of these characteristics, however, are of recent development, especially in Western Europe. Many fascist and other repressive regimes are recent memories in many parts of Europe. Conflicts over nationalism and national identity are still widespread. The appearance of the affluent middle classes is largely a product of the period since 1945.

State building is a brutal and disruptive process. Only after centuries of resistance and struggle did representative political institutions develop. These are still confined to only a minority of states. In Western Europe the institutionalization of political parties, party competition, and representative politics in general developed only under conditions of persistent economic growth, domestic welfarism, and dominant position within the world-economy. Many of the parties, especially the Socialist, Communist, and Christian Democratic mass parties, were the product of earlier struggles for extending the right to vote and "taking over" the state in the interests of one group or another (Duverger 1959). Their contemporary organization and objectives can only be understood in terms of this historic struggle with the state.

In general, political parties, whatever the system of parties (single party, dual party, or multiparty), have two reasons for existence (Burnham 1984). From a "bottom-up" perspective, parties provide a means for organizing and mobilizing a population and articulating a wide range of popular demands. National interest groups fail to do this because they are skewed towards the narrow interests of those with sufficient lobbying resources. Parties are fundamentally involved in redistribution activities rather than merely reproducing the status quo (Burnham 1984: 130–1). When there are two parties, however, at least one of them will portray itself as "the party of the state." From a "top-down" perspective, therefore, parties provide state-building elites and the state itself with political legitimacy. Conflicts are channeled and tamed; political stability provides crucial support for activities of capital accumulation and political control. The costs of legitimacy, however, often exceed the rewards. There have been numerous situations in which this has occurred. Elites then act to shield or insulate themselves from redistributive pressures. The forms the action takes are remarkably diverse. They have varied according to the specific histories of particular states as well as in terms of the historical epoch and relationship to the world-economy when insulation became necessary. They include oligarchy or rule by notables, military or royal–military junta governments, and single-party regimes (Burnham 1984: 117–28).

In the process of state building, not only political parties but also systems for organizing local government and the territorial organization of the state become necessary. Linking the center, or dominant place, to the places incorporated by the state has been in Tarrow's (1977: 47) words: "a fundamental problem in the development of the modern state in the West." And elsewhere too, one imagines. The linking process has had three major aspects: "the envelopment and control of formerly autonomous jurisdictions; the unification of previously independent ethnic and linguistic regions and creation of uniform legal codes; and the incorporation of the grassroots public into participant roles in the national system" (Tarrow 1977: 47). However, this has been organized in different ways in different states. In some cases, as in Switzerland, local territorial units have been given or retained considerable power in terms of initiative (regulating and legislating in their own interests) and immunity (acting without fear of intervention by higher tiers of the state). Elsewhere, for example, in France, there is an absence of both elements of local autonomy, or, as in Britain, initiative but no immunity. In the United States there is a very powerful ideology of autonomous local government, but, as Clark (1984: 205) suggests in his seminal article on local autonomy, the reality "seems much closer to absolutely no autonomy." Of course, this point is debatable, particularly when the American case is compared to that of France or Britain. "Ideal types" of local autonomy, as of anything else, have limits when examining historical cases.

Certainly the power of the state is everywhere very great when compared to that of local political units. What is less clear, however, is whether this is a zero-sum relationship. For example, more direct administrative relationships between local and national levels can lead to a greater sharing of power and resources between levels than when they are more isolated from one another. Suttles (1972 : 258) points to this "paradox of centralization" when he writes:

> Are, for example, Stockholm and its new suburbs like Farsta and Vällingby less powerful and distinct because the Swedish prime minister appoints the city's mayor and the new suburbs have been totally planned and built by the central state? In fact, the direct relationship between the mayor and the prime minister may establish a line of administrative authority in which power and responsibility can be joined.

The central state itself also *depends* upon attracting support and legitimacy from large sections of the territory over which it holds sway. The successful and continuing integration of the modern territorial state requires a considerable degree of popular consent and participation in political life. The genius of the Italian Marxist, Antonio Gramsci, lay in pointing this out. In particular, Gramsci argued that the dominance of a state-building elite rests upon the perpetuation of a "hegemony" or cultural form through which its dominance is defined as legitimate (MacLaughlin & Agnew 1986). Though the hegemony of the elite defines the limits of "what is seen to be possible" and limits the growth of alternative horizons, there is nothing absolute, determined, or inevitable about it. It is sustained only with great difficulty and through control over major educational, cultural, and political institutions.

Of major importance: under conditions of economic and political inequality between places, a hegemony can only be maintained by pursuing policies with geographical consequences and place-biased policies that favor those people and places that favor the hegemony. This is most obvious in the case of states such as Italy that have been integrated in part through clientelistic rather than mass or party politics. As Tarrow (1977 : 63) notes:

> Every political system develops mechanisms to link center and periphery, national and local levels of government. In the absence of a strong state, a standardized culture and language, and a legitimizing national myth, the Italian system was structured around the network of personal and political relationships that developed between local notables and the central state, frequently through Parliament, but increasingly through a network of personal and family ties in the bureaucracy. For Italian political genius lies less in the creation of new administrative structures or of symbols than in the sharing out of public goods among a host of private claimants.

Arlacchi (1983:22) puts it another way but the message is much the same: "clientelism filled the space between the new alien administrative machinery of the state and the existing community." In turn, of course, this weakened the absolute power of the state at the same time that it increased the state's dependence on local sources of power.

But not only states characterized by clientelism exhibit a "geography of hegemony." As Dulong (1978) argues, in states in which mass (or party) rather than clientelistic politics dominates, some places represent "success" and others "checks" for the *stratégie hégémonique*. A symbiotic but volatile relationship exists between political parties and their supporters. Political performance is measured by success in resolving the problems and ideological dilemmas faced by supporters. When supporters are concentrated geographically, political performance is in large measure viewed as geographically specific. MacLaughlin and Agnew (1986) attempt to demonstrate how this argument fits the implementation of regional industrial policy in Northern Ireland in the period 1945–72. Bensel (1984) provides considerable evidence which suggests that American congressional politics is largely about introducing policies that benefit geographical constituencies which in turn give their support to particular parties and specific candidates.

The other side of the coin, so to speak, is that places which are nonsupportive or in the minority suffer the consequences. Often this is justified on "rational" economic grounds (Agnew 1984b). These other places have the wrong population mix, they are too far away from "markets," or they are "in decline." There are groups of intellectuals committed to providing such rationalizations. In recent years in the United States, the economic decline of certain places has been met by proposals for abandoning the places experiencing it, e.g. inner-city neighborhoods, the entire northeast region, New York City (Winnick 1966, Report of the President's Commission 1981). While this does of course reflect an analysis of changing global economic conditions, it also reflects a changing geographical balance of power within the US. Some places are expendable because political power and the balance of representation have shifted in favor of other places.

Places, therefore, are connected to the state through its organization into various tiers of administration and the geography of its hegemony. The state survives and prospers as long as it can hold together the territorial coalition of places that gives it geographical form. The state is dependent on places for support as they are dependent on it for political influence. The political parties and local–central links it provides a locus for are in turn their main channels of political expression.

It should be clear by now, however, that states can be expected to differ, one from another and individually over time, in the extent to which a place-based political life is possible, has been stimulated, or is significant. First, states in which coercion rather than consent is the rule do not allow for place-based political expression. Only when such coercion is removed, as in Italy after Fascism or Germany after Nazism, and a place-based expression returns can the artificiality of nationalization under coercive regimes be revealed. It is also significant that under extremely coercive regimes, surveillance is always organized on a geographical basis. In addition to practical considerations, local solidarity is seen as a major threat.

Secondly, some systems of local government, some political party systems, some electoral systems, and some experiences of political integration, particularly those encouraging territorial commitments in political outlook, encourage a place-based political life. Casual empiricism suggests, for example, that among contemporary states Italy provides more opportunities for place-based politics than does Britain (Brusa 1984b, Sharpe 1982).

Thirdly, and finally, the histories of state formation and world economic development suggest that place-based political expression varies in visibility over time. In particular, it was most pronounced in the early period of state building but declined in the early 20th century. At that time, in both Europe and the United States, the argument goes, a "nationalization" of political life in the sense of a diminution of between-place differences in interests and outlooks was paralleled by a rise in a politics of nation or class (Schattschneider 1960). Recent economic stagnation and disillusionment with national political institutions, however, have given rise to a reinvigorated politics of place (e.g., Burnham 1982, Berrington 1984). Certainly, studying place and politics, or at least recognizing a role for place in understanding political activities, has come in "waves" that parallel the ups and downs of "nationalization." But the fact that place has re-emerged, substantively and intellectually, suggests that it was also "there" all the time, only structuring social relations that were then more similar across places than both before and since. Place is important, therefore, even when local *outcomes* are not distinctive. That is at the core of the structuration thesis and the central claim of this book.

Why adopt the place perspective?

But why should anyone choose to adopt this approach in preference to the one they now take? I think there are a set of advantages to this approach that should be confronted before returning to "business as usual." These are discussed in terms of certain critical debates in contemporary social science.

The first of these is the advantage that lies in grounding the categories of social science in the concrete analysis of everyday life. Class, for example, is often reified as a thing or a universal geist that moves people to act in certain ways. Williams (1977:80–1) takes issue with this conventional view: "The analytic categories, as so often in idealist thought, have, almost unnoticed, become substantive descriptions, which then take habitual priority over the whole process to which, as analytic categories, they are attempting to speak." But in structuration theory, class is a historical or relational category. The self-consciousness of groups is emphasized. Thus class becomes "an active process, which owes as much to agency as to conditioning" (Thompson 1972:9). The presence or absence of class, its reality as an empirical object, becomes a matter for historical and place-specific analysis rather than an a priori theory (Thompson 1978).

The same goes for the categories of religious affiliation. The term "Catholic," for example, differs in content and meaning depending on the variable experience of specific groups of Catholics. The Catholicism of northwest France, for example, is different from that in the Nord or the Pas de Calais (Goguel 1951a, LeBras 1949). Of course, given their denominationalism, this is doubly true of Protestants.

A second advantage is the substitute this approach provides for the flawed project of positivist social science. Positivism has as its central tenet the search for empirical laws independent of time and space from which causes can be inferred. The difficulty and futility of this task are well documented (Szymanski & Agnew 1981, Gregory 1978, Sayer 1984). Alternatives are legion, but, as argued earlier, structuration theory provides the only base upon which to combine an emphasis on agency with a continued commitment to causation. This approach allows for the historical specificity and uniqueness of places while proposing that these "multiple outcomes," if you will, are the product of a "one to many correspondence" (Laslett 1980:219) between cause and effect. Thus the structuration of social relations in everyday life contains many similar elements from place to place (e.g., class relations, central–local government relations, etc.), but produces many different outcomes in different places.

Other advantages derive from resolving problems central to orthodox and antithetical political sociology. One, of course, is an approach to resolution of the structure–agency problem. The approach proposed here recognizes human action as both motivated and intended but, at the same time, as both mediated by social structure and generative of it. This is not an easy task. In practice there is a strong tendency to slip towards either voluntarism or determinism. But this approach at least holds out the possibility of resolving this persistent dilemma in social science.

Another advantage lies in the denial of what can be called "a priori evolutionism." A conspicuous feature of Western social science in general and political sociology in particular has been a predilection for dividing up history into evolutionary stages. This has led into difficulties when, for example, features of one stage persist into another, or when stages are bypassed or fail to appear. Twentieth-century social science has become a graveyard of 19th-century evolutionary schemas. Unfortunately the dead continue to haunt the living.

A final advantage is particularly appealing to those disturbed by the

economic determinism of much, particularly "antithetical," political sociology. This lies in the potential of the structuration perspective for resolving
"the base–superstructure problem" that lies at the root of much economic
determinism. Cultural phenomena – ways of life, systems of meaning and
other "noneconomic" phenomena – are often seen as "reflections" of an
economic base, serving the function of "reproduction" for its survival.
Recent work suggests that this type of reasoning is fundamentally flawed
(Brenner 1977, Williams 1977, 1980, Wood 1981). People are *practical* rather
than economic "in nature," the alternative argument goes. The economic
signifies one set of practices, meanings, and values which, though dominant
in certain societies, are practical rather than transcendental in origin. Thus
the practical nature of everyday life rather than the abstracted nature of
economic organization is the critical nexus for explaining social organization.

Conclusion

A variety of critical dilemmas faces contemporary social science. As yet these
have not been faced squarely in political sociology. Indeed, current practice
seems largely oblivious to them except insofar as existing approaches can be
"patched up" on an ad hoc basis (see Ch. 6). But the "microsociological
challenge" to both macrosociology and individualism needs to be taken up.
This has been the burden of this chapter: to lay out a rationale for political
sociology that responds to the challenge. The main focus has been on
identifying the major definitional elements of place – locale, location, and
sense of place – and grounding these in structurational sociology. Above all,
place is defined as the geographical context or locality in which agency
interpellates social structure. Consequently, political behavior is viewed as
the product of agency as structured by the historically constituted social
contexts in which people live their lives – in a word, places.

4 Place and political behavior

Behind much contemporary political sociology lies an exceedingly simple image of political alignments and change. It is often assumed that political behavior is based on class or other universal categories, with changeability a function of the passing issues of national politics. However, as Butler and Stokes (1969 : 135) have commented, "any idea that class provides a universal ground for party allegiance is difficult to reconcile with the extraordinarily uneven spread of party support between [British] regions, continuing decade after decade." And this is not just a result of differences in the class composition of regions (p. 137). So if in Britain – where, to quote a famous aphorism, "class is the basis of British party politics: all else is embellishment and detail" (Pulzer 1967 : 98) – the effects of local political environments are apparent, what might we expect elsewhere? Britain is the paradigm case of the "modern' nation in which national, functional cleavages *should* have transcended all others (Lipset & Rokkan 1967).

A major implication of structuration theory is that the conventional way of thinking about social cleavages and political behavior will not do at all. Rather than seeing local variations as deviations from a national norm, the national norm should be viewed as constituted by locality-based structuration. This is *not* to say that national political parties, national administration, and large-scale economic and political institutions are irrelevant to political behavior: it is to argue that insofar as they intersect with "paths and projects" relevant to political behavior, these structures are reconstituted and take on meaning in place through ongoing social and political practices.

The particular relevance of place to political behavior lies in its directing attention to the concrete social–geographical bases of political outlooks and activities. In short, the argument is that place-specific social structures and patterns of social interaction give rise to specific patterns of political behavior. "Active socialization" in place produces particular political outcomes. To the extent that these outcomes are similar across places, one can talk of "types" of place. But it is in specific places that the causes of political behavior, if you will, are to be found. The burden of the rest of this chapter is to demonstrate the truth in this position.

Empirical themes

The possibilities for a place-based political sociology can be demonstrated by addressing a set of themes drawn from the confluence of conventional political sociology and the specific orientations of the place-based perspective. First is a focus on the historical constitution of political behavior in places, using historical case studies. Second is an argument for the contextual sources of political behavior. Third is an attempt to distinguish "types" of place in terms of aggregate political behavior. Fourth is an attempt to identify the place-based practical roots of changes in political behavior. Fifth is an approach to explaining the rise of new political movements in terms of place-based structuration. Sixth is a brief survey of some recent literature that links "policy outputs" to local political environments. Seventh, and finally, is an examination of the degree to which places are "politicized" or "depoliticized" in terms of the level of political activity and integration or abstention from current political institutions and outlooks.

Each of these themes has received some attention previously. Existing work, primarily in social history and French electoral sociology but also in the "contextualist" and "group-ecological" traditions of Anglo-American voting studies, provides considerable conceptual and empirical support to the place perspective. Though it is not feasible to cover this literature in an encyclopedic manner, it is possible to draw on exemplary analyses that fit specific themes, or in some cases cover several. Some studies cited did not consciously pursue a place perspective but nevertheless provide evidence that supports such a viewpoint. This reflects a state of affairs in which most work that fits under the rubric of political sociology has not until recently seriously entertained the virtue of a place perspective even when empirical evidence might support it. Verily, facts do not speak for themselves!

THE HISTORICAL CONSTITUTION OF POLITICAL BEHAVIOR IN PLACE

The term "political behavior" covers a range of activities – everything from lobbying and pressuring political institutions through organizing or participating in protests, political strikes, and demonstrations to voting, canvassing for political parties, wearing party "buttons," and so on. Even "not voting", or abstention from other political acts, can be seen as political acts (Lancelot 1968, Burnham 1982). Typically, political sociologists have focused on voting rather than other activities. Fortunately, some others, particularly historians, have taken a more catholic approach. There are examples of each emphasis, however, which indicate the elements involved in reconstructing the historical constitution of political behavior in specific places.

Some of the best examples of an "implicitly" structuration approach to political behavior are the works of certain social historians. Joyce's (1975, 1980) research on factory politics in late-19th-century Blackburn, Judt's (1975, 1979) investigation of early-20th-century politics in the Var (France), and Merriman's (1979) study of political conflict in late-19th-century Limoges are exemplary.

Joyce demonstrates that the political situation in localities in late-19th-century Lancashire, in particular in Blackburn, was closely related to "individual town character" through the nature of factory ownership and the local structure of labor. This approach, as he puts it (Joyce 1975:527),

> confronts the orthodoxy, still lively, that has the nineteenth century progressing inexorably to an individualistic democracy in which public opinion, separable from influence, achieves its free play in the later century, the 'national' triumphing over the 'local', the deference community going to the wall.

In late-19th-century Lancashire, voting was dominated by the factory–communal nature of local life. Studying voting returns for the 1868 general election, Joyce shows a distinctive correlation between the political affiliations of cotton-mill owners and the votes of their employees (Joyce 1975:536–41). He argues (p. 539) that "some combination of dependence, obligation and habituation" was at work in producing the pattern. Coercion in its various forms – actual and potential dismissal from work, intimidation, and so on – while present, does not constitute a sufficient explanation. Rather, political allegiance was "based essentially on consent; though a consent that was part of a paternalist exchange to which authority was integral" (p. 542).

Employer influence is traced through a wide variety of local activities and local politicians. The factory penetrated deeply into political life through financial and job-related rewards for political loyalty, employer patronage of Sunday schools, sponsorship of political clubs, and control over churches and chapels (p. 549). The substance of this involvement varied, depending upon whether employers were Tory or Liberal. But whatever the political persuasion, localities were equally distinctive and cohesive:

> Distinctive employer areas studded all Lancashire towns. A brief catalogue must serve to show how typical Blackburn and Bury were: Ashton's West End, Buckley's Dukinfield, Platt in Oldham (the largest engineering employer in the world for some of the period), "Nat" Eckersley and the Earls of Crawford in Wigan ("The good old town is but an appendage of the House of Haigh"), Glossop, Edward Hermon of the massive Horrocks's of Preston, the Birley sway in Manchester. In small towns a single family could dominate – the Ashtons in Hyde, Pilkingtons of St. Helens, Garnetts of Clitheroe, and so on (Joyce 1975:543).

Towards the end of the century, the factory basis to popular politics diminished. There was a rising chorus of "oppositional" demands to the hegemony of both Tory and Liberal employers (Joyce 1980). This reflected changes in the structure of the cotton industry, particularly its increasing concentration of ownership, and the opening up of localities to the discourse of "class politics." The attractiveness of this new discourse, however, was based largely upon its meaningfulness in explaining local shifts in work authority and economic dependence (see Jones 1983: 21–2).

Judt's (1975, 1979) investigation of the development of socialism in the department of Var challenges those "facile" accounts of working-class socialism predicted upon the "immanence" of socialism among the new industrial proletariat (Judt 1975: 76–7). Judt shows how the "very socialism of the urban milieu" became planted in the countryside of the Var. He does this by identifying "structural," "circumstantial," and "traditional senti-ment" components of a contextual explanation.

Judt notes, for example, the coincidence between those areas producing for a wider market (e.g., the wine-producing communes) and those overwhelm-ingly socialist in their votes (Judt 1975: 67). He argues that this "coinci-dence" was the result of the transformation of the poorer, small proprietors into a rural proletariat; or, more likely, that the fear amongst such proprietors of this transformation pushed them towards socialism. Other "structural" factors, such as whether or not settlements were closely agglomerated (the socialist vote was much higher in denser, more compact settlements); and the level of education and literacy (support for socialist candidates was much higher where educational and literacy rates were higher), provide additional elements in his contextual explanation (pp. 71–2).

"Personal" and "circumstantial" factors include the deep interest taken in the Var by prominent socialists from Blanqui to Guesde and Vigne; and the economic depression of the 1880s and 1890s, which affected Provence and the Var much more than the semiautarkic economies of other parts of rural France (Judt 1975: 74–6).

Finally, Judt's category of "traditional sentiment" concerns the "collective memory" of people in those localities in which socialism took hold. Socialist support was party-specific and thus not just one manifestation of a vague "Leftism" that some commentators (e.g., Goguel 1951b) have ascribed to certain areas, such as the Midi. But a 'tradition' of socialism was grounded in specific localities in the Var by the *sociabilité méridionale* of everyday life: the propensity to conduct political, economic, and personal business in public (Judt 1975: 65). Rather than the product of a nebulous intergenera-tional transfer of ideology, therefore, socialism in the Var was grounded in a *persisting* set of institutions that were collectivist and populist in nature (see also Agulhon 1983).

What all this adds up to is an argument that there is no necessary conflict between elements of "structural," "traditional," and "circumstantial" accounts of the development of socialism in the Var. On the contrary, when combined in a contextual account, they complement one another. Judt (1975: 82) concludes:

At a certain moment, the economic, the geographic, the social bases of Provencal society were drawn into a chance meeting with a new political idea; the two were not of necessity interlinked; they became so through the events of the period, and over time. What was then born was a political tradition which, following a certain time-lapse, could no longer be accounted for in structural terms at all – anyone attempting to account for the political behavior of the Var through analysis of its present social structure would be understandably baffled. History, as it were, has taken over from political science.

Merriman (1979) provides an account of political change in the city of Limoges in which the city itself – "its configuration, neighborhoods, and relationship to its region" (p. 131) – is stressed as the most important constituent or focus of explanation. Merriman identifies the butchers of Limoges as symbolic of the old city and the porcelain workers as the symbol of the new city that was emerging in the late 19th century. He interprets the history of the city largely in terms of the conflicts and changing balance of power between these two groups.

Amongst the workers, established city dwellers (*villands*) and recent immigrants (*bicanards*) shared common work experience and residential areas. These were instrumental in the growth of a shared consciousness as workers. But this was reinforced by their common geographic origins and common dialect. The butchers, by way of contrast, were united by allegiance to their trade, "their neighborhood, saints, and to a hierarchical and vertical system of political and moral authority based upon the power of father, priest and king" (Merriman 1979:138). Different solidarities, allegiances, and interests, defined by different work experiences and different but equally strong senses of neighborhood, therefore, produced the peculiar combination of majority socialism and minority monarchism that characterized late 19th-century Limoges (see also Merriman 1981, Margadant 1981).

These three examples involving different places in the late 19th and early part of this century pinpoint the specific attributes of a historical–contextual approach to understanding political behavior. The setting is fundamental – the intersection of established structures and evolving practices provides the explanatory locus – however the empirical materials are sorted and arranged. Of course, one reaction could well be: this is all very well, but it refers to a long time ago. Blackburn, the Var, and Limoges are no longer distinctive places. What is the evidence *against* this view?

PLACES AS CONTEXTS FOR POLITICAL BEHAVIOR

One tradition in political sociology, beginning perhaps with Tingsten's (1937) research on election statistics, has also situated *modern* political behavior in a place-based framework. Tingsten (1937) demonstrated that the level of participation of working-class voters in elections was a direct function of the proportion of workers in the local population. He also showed that the more workers there were in an electorate, the greater was the size of the vote for "working-class party" candidates (Tingsten 1937:126–7, 156). The historically changing composition of an electorate was therefore taken to indicate a social context in which the likelihood of voting a particular way was reinforced by the presence of similar or like-minded people.

This "social influence" approach has given rise to a substantial literature, although "behavioral contagion" has increasingly substituted for a primitive version of structuration. Contextual models have focused on interpersonal communication and influence and so-called "neighborhood effects" rather than on the social structuration of political behavior (Cox 1969a, 1972; Przeworski 1974; Wright 1977; Huckfeldt 1979). In particular, they often hypothesize about the conversion of "minority" group individuals to the partisan preferences of the majority group in a constituency. They thus *add* a contextual or local "residual" effect to a national structural model (see Ch. 6). Despite this separation of the individual from the social, there is abundant evidence that social contexts *do* count in explaining both political stability and political volatility (Huckfeldt 1983). But this is not just in terms of local information flows, as advocates of neighborhood effects propose, but also in terms of local residential and workforce environments (Epstein 1979). It is interaction patterns and their settings *in toto* that structure partisan and ideological attachments (Putnam 1966, Ennis 1962, Wright 1977, Andrews 1974, Percheron 1982). This literature reinforces the conclusions of the more historically – and place- – specific studies outlined previously. Political science, as it were, speaks to history!

TYPES OF PLACE AND POLITICAL BEHAVIOR
The major contribution of "classic" French electoral sociology to a place-based political sociology has been its focus on classifying places and regions in terms of political expression, and then explaining the classification in terms of place-related characteristics (Derivry & Dogan 1986). Siegfried (1913), the founding father of this approach, adopted the position that determining factors, from the geology of an area to the religious preferences of its population, could be identified and given the status of "general laws" (Morazé 1949, Aron 1955, Berger 1972, Coulin 1978). Later proponents of the approach abandoned this objective and chose to focus on the *intersection* of "local" and "external" causes (Agulhon 1950, Goguel 1951a, Klatzmann 1957). Types of place and their associated political behavior rather than the identification of transcendental general factors became the major concern of the electoral sociologist (Goguel 1951b).

In the 1950s and early 1960s this position was a very influential one in French political science. It is difficult to find items in the *Cahiers de la Fondation Nationale des Sciences Politiques* from this period that are not influenced by Goguel and his followers (e.g., Goguel 1954, Fauvet &

Mendras 1958, Goguel 1962, Marie 1965, Lancelot 1968). Since 1968 the influence has been much reduced, displaced by shifts towards either psychological reasoning or abstract sociological theorizing (e.g., Leca 1982, Lévy 1984a).

There has been a growing interest in the English-speaking world and elsewhere in "ecological" models of voting patterns. This represents increased confidence in the statistical methods upon which ecological analyses are based, after many years of worrying about the "ecological fallacy," or the danger of making inferences about the behavior of individuals from aggregate data. It also reflects declining confidence in the aphorism that a census-defined "class" category is the single basis for party politics in countries such as Britain or the United States. Other factors, such as the increased availability of socioeconomic data by constituency units and the growing interaction between sociology and political science, may also have been important in fostering interest in the ecological models (Taylor & Johnston 1979: 207–10, Lipset & Rokkan 1967, Dogan & Rokkan 1969).

A good example of an ecological analysis that results in a classification of places according to political behavior is that of Crewe and Payne (1971, 1976). They defined a simple regression model in which the percentage of the Labour vote in a British constituency was related to the percentage of manual workers. Residuals from this regression were divided into two distinct types, indicating respectively the types of place in which Labour did better or worse than expected. Labour did poorly in agricultural constituencies and much better in heavily working-class constituencies such as mining areas. The fine level of variation picked up by this model is illustrated by the identification of specific constituencies in which local influences are predominant – for example, the local influence of Enoch Powell and the issue of immigration in the West Midlands conurbation.

In their first paper, Crewe and Payne (1971) focused on the top fifty positive and negative residuals, but in a later paper they moved beyond the classification of residuals to a multivariate analysis of them (Crewe & Payne 1976). They expressed their final model in the following form (p. 64):

The Labour share of the combined Labour and Conservative vote in any British constituency in the 1970 election = 30.7% + 0.24 × the constituency's percent of manual workers − 4.5 percent if the seat is agricultural + 23.0 percent if the seat was 'very' Labour in 1966 + 9.7 percent if the seat was 'fairly' Labour in 1966 − 7.3 percent if the seat was 'fairly' Conservative in 1966 − 16.2 percent if the seat was 'very' Conservative in 1966 − 3.3 percent if a nationalist contested the seat in 1970 + 3.6 percent if the seat is a 'mining' one + 2.3 percent if the Conservatives lost in 1966 and captured less than 60 percent of the total vote of the losing parties, − 6.4 percent if Labour lost in 1966 and captured less than 60 percent of the total vote of the losing parties.

Residuals from this model were also examined. They provide the basis for investigating "local effects" left after national-scale effects in places have been accounted for.

A different approach might resist the temptation to divide the causes of place-based political expression into the national and the local, as if they can be separated from the historical constitution of a place itself. The precise meanings of such concepts as "manual workers" and "agricultural" are clearly time- and place-specific (Taylor 1971). This, of course, reduces the possibilities of statistical analysis using fixed national categories with fixed meanings. From the perspective of a place-based political sociology, it probably leads us back to the approach of the French electoral sociologists as superior to, if "softer" than, that of the electoral ecologists in linking types of place with political behavior. The best way, of course, would be to combine ecological analysis with the sort of research described in the previous two sections (Derivry & Dogan 1986).

PLACE AND CHANGES IN POLITICAL BEHAVIOR

The image of political behavior in conventional political sociology is that of fairly persistent social cleavages generating fairly predictable political out-looks and actions. Changes in behavior, such as shifting party support or abstaining, are seen as the product of temporary national issues, such as the state of the economy or specific foreign policy issues. After a while equilibrium will return; the fundamental social cleavages, particularly those of a functional or nonterritorial nature, will reassert themselves (e.g., Lipset & Rokkan 1967).

There is a tradition of research, however, that takes political change more seriously, and links it to historical shifts in place-specific economic interests and political sentiments. This is the "group-ecological" tradition of American political science. Perhaps Turner, the "frontier theorist," can be thought of as the founder of this school (Archer & Taylor 1981), but in recent times its most persuasive advocate has been Key (1949, 1974). Key was interested in elections insofar as they were part of a process of distributing power within a national society, so he emphasized the mobilization of individuals in interest groups and political parties rather than individual voting decisions. The historical–geographical structuring of the American economy was identified by Key as the major process whereby the sorting out of people into groups took place.

Many popular accounts of American elections have followed the trail pioneered by Key (e.g., Phillips 1969). They also focus on one of Key's best-known concepts, that of the "critical election," in attempting to explain regional realignments in support for political parties. Critical elections are ones that mark rapid and semipermanent shifts in partisan alignments as new alliances and coalitions form in response to crises or new issues (Key 1955:5). The slavery issue in the 1850s and the economic depressions of the

1890s and 1930s are identified as specific critical periods in American electoral history. But rather than viewing these as *national* events, Key tied them to interests and sentiments that were place-specific.

Of contemporary observers of American politics, Burnham (1982) is the one most closely identified with this approach. He divides recent American political history into three periods (Burnham 1982:17–18). The first of these, the "system of 1896," revolved around a "center–periphery" struggle. The Republican party became the dominant party in the socioeconomically "developed" parts of the United States. The underdeveloped South became the stronghold of the Democratic party. The "colonial" West was a zone of contestation but, as Burnham (p. 18) puts it, it was "episodically insurgent against domination by Eastern capital and its hegemonic Republican political instrument."

The second phase, the New Deal revival of party and mass electoral participation, came on the heels of the Great Depression and its shattering of the "core" economy. This phase lasted from 1932 until about 1960. The Democratic party established itself firmly in the industrial belt of the United States, particularly among new immigrants, and with its Southern wing emerged as the dominant electoral coalition.

The third phase, beginning in the 1960s, marks the breakdown of the Democratic coalition with the emergence of divisive foreign policy issues, the recurrent crises of the national economy, and the "rise" of the West and the South as bastions of right-wing politics. One important feature of this period has been the increase in negative attitudes towards political institutions and leaders and the decay of electoral participation among the working class.

Burnham examines these historical phases in terms of local trends. A central theme in his discussion of the third phase is "the extent to which the imperatives of the class–ethnic New Deal realignment have been relevant to the local social structure and political culture" (Burnham 1982:44). In the case of Ohio, for example, Burnham draws on Fenton's (1962) discussion of the 1962 governor's election to discuss the intersection between local settings and political behavior. Basic to Ohio's social structure, Fenton argues, is a wide diffusion of its working-class population among a large number of medium-size cities and small towns. Resting on this diffusion is the weakness of labor unions and a chronically disorganized state Democratic party. Ohio also lacks organizations that report on politics from a working-class point of view. The upshot of all this is that "to [a] much greater extent than in other industrial states, potential recruits for a cohesive and reasonably well-organized Democratic Party in Ohio live in an isolated, atomized social milieu. Consequently they tend to vote in a heavily personalist, issueless way, as the middle and upper classes do not" (Burnham 1982:40). In the absence of local political organizations, issues, and candidates around which an intense polarization of voters could develop, the New Deal system failed to incorporate large numbers of places. Elsewhere it broke down as it failed to match national political promises to local realities.

PLACE AND NEW POLITICAL MOVEMENTS

Most popular political movements and parties, even those wanting to operate at a national level, have their origins in specific places or regions. The clearest cases are of regionalist or separatist movements. Much analysis of such movements regards their origins as lying in the conjunction of a spatial division of labor with cultural distinctiveness (e.g., Hechter 1974, Gourevitch 1979, Rokkan & Urwin 1983), tracing their emergence in the 1960s and 1970s to the relative economic decline of previously dominant "core areas". In turn the geographic division between core and ethnic periphery became more salient relative to functional cleavages such as class. The attractiveness of the core to peripheral elites diminished. New movements were founded by the disaffected and appealed to the periphery's population.

The actual process may be more complex than this. Indeed, *within* peripheries there can be considerable differentiation of support for new movements and parties (Balsom 1979, McAllister 1983, Agnew 1984a). Or there can be divisions within movements, with segments concentrating their support in specific areas (Rumpf & Hepburn 1977, Fitzpatrick 1978, Carty 1981). High levels of support or movement strongholds may be limited to only certain types of place, ones in which class solidarities have either failed to develop or have been effectively displaced by territorial–ethnic ones (Gourevitch 1979, Agnew 1984a).

The connection between place and new political movements is, however, not only characteristic of obviously regionalist movements. It is well known that the British Labour party originated as a mass movement in the East End of London, West Central Scotland, and South Wales (Pelling 1967). Less well-known are the place-specific origins of Italian Fascism and German National Socialism, paradigmatic cases of "national" political movements.

There has been a recent resurgence of interest in the origins and rise to power of Italian Fascism (Granata 1977, 1980; Corner 1975; Vivarelli 1979; Cardoza 1982; Bell 1984). A major thrust of this literature has been examining the connection between local "Fascisms" and the emergence of a national "Fascism." The traditional view was to see Fascism as a city-based movement in its early phase but challenged by an increasingly powerful rural variety, a "reactionary" Fascism seen most clearly in the Po Valley and especially in the provinces of Ferrara and Bologna. This latter variety eventually replaced its urban rival in national importance (Bell 1984:2). Recent research suggests, however, that prior to the general strike of August 1922 there was no "classical" model of Fascism, urban or rural, but only a variety of local "Fascisms" (Granata 1980). Granata finds similarities to the rural Fascism of the Po Valley in other regions, most notably Tuscany, Umbria, Apulia, and parts of Piedmont. But other rural areas of the north and center were not hospitable; Southern Italy was, with the exception of Apulia, entirely without "Fascisms" (Granata 1980:513–29). Large cities and smaller manufacturing centers contained distinctive Fascist movements that had to cope with working-class opposition and popular resistance to their activities (pp. 529–30).

In a study of one manufacturing town, Sesto San Giovanni, Bell (1984) explores the intersection between the development of working-class culture in the town and the appearance of a local Fascist section. He shows that links between the Sesto *fascio* and the national headquarters in Milan, only a short distance away, were tenuous at best. "Fascism in Sesto was essentially a local and semi-autonomous movement" (Bell 1984:11). Many of the leading Fascists were local professionals, small local manufacturers, and members of the pre-World War I political elite. Bell emphasizes the "threat" that the electoral success of the local Socialists posed to these elements in the population. He then details their attempts to build a local mass movement by drawing on the experience and example of working-class organizations. In particular, they were able to exploit schisms between working-class organizations, especially the Socialists and the Communists, to present themselves as a viable alternative (Bell 1984:23).

An important conclusion of research such as that of Bell and Granata, therefore, is that Italian Fascism was not a unified movement or bloc but a collection of heterogeneous and diverse segments rooted in different types of place. Rural Fascism in the Po Valley was, as Lanzillo (1922:226) expressed it, a "cruel and implacable movement of interests, which devoted itself to well-defined goals." In particular, the perpetuation of landlord power against peasants and *braccianti* (landless laborers), and removing agrarian trade unions and cooperatives were important and clearly stated objectives. In an industrial town like Sesto, however, Fascism was an attempt in a hostile environment "to reestablish the sort of conservative political solution that had prevailed before the war, while at the same time devising some new strategies or borrowing strategies from other movements in an effort to gain popular support" (Bell 1984:11).

The rise of the German National Socialist party in the early 1930s has, if anything, generated more controversy among students of new political movements than any other single topic. The consensus view has been that the Nazi electorate was predominantly lower middle class, with very few votes coming from blue-collar workers or the upper strata of Weimar society (Bullock 1964, Moore 1966, Kornhauser 1959, Lipset 1960). But this viewpoint was never examined in anything other than a cursory way (Hamilton 1982:9–36). Hamilton provides evidence that support for the Nazis in the 1932 election came from two distinctive types of place: the upper middle-class suburbs of northern German cities such as Berlin and Hamburg, and the Protestant countryside of central and northern Germany. In the Catholic countryside and Catholic cities such as Cologne and Munich, the

Nazis gained relatively few votes. In those settings the *Zentrum* party (moderate Catholic party) was well entrenched, and able to counter the Nazis' appeal (see also Passcher 1980).

Stone (1982) identifies religious mix as an important intervening variable in the success of the Nazis. He argues that Protestants were particularly pro-Nazi in mixed districts such as Franconia, southwestern Germany, and Westphalia. In Westphalia, he says (p. 26), "the Protestant and Catholic villages, although neighboring, were voting overwhelmingly Nazi and Center [*Zentrum*] respectively." By way of explanation he suggests that attitudes towards debt were fundamental to this geographical differenti-ation. "Protestants took on debts and modernized their farms. In 1931 they faced interest charges which almost wrecked them at a time of agrarian slump." Catholics, however, avoided debt and farm modernization. They were thus less vulnerable to the Nazi appeal, including its anti-Semitism. They were also, especially in the mixed districts, much better than the Protestants at creating political machines (Blackbourn 1980). This further reduced their electoral vulnerability to the Nazis.

That the Nazis faced different audiences in different settings is clear. Consequently they organized differently and pushed different issues as the need required. To Hamilton this was the secret of Nazi success. They con-ducted a "differentiated struggle" depending on the character of a place and the nature of their opponents. Their most decisive action, according to Hamilton was their "move to the countryside" in the period 1928–32. The Protestant countryside provided "a largely untouched segment of the population; it also put the party in contact with a segment whose party ties were very loose" (Hamilton 1982:363). In the towns and cities, the Nazi message – largely one of anti-Marxism rather than the anti-Semitism favored in the countryside – struck a responsive chord most of all in the "best districts." Hamilton comments (p. 421): "In *no* city did the results show the expected pattern of a pronounced and distinctive National Social-ist tendency in the mixed districts, those with large lower-middle class populations."

Rather than obtaining support from a single functionally distinct segment of the population, therefore, the National Socialists appealed for support successfully in some settings and received little or none elsewhere. Hamilton proposes (pp. 437–42) that this conclusion supports the view that for success, new movements such as National Socialism must combine an appeal to place-based interests (for example, indebtedness) and sentiments (especially nationalism in this case) with an "injection of new cadres" who argue for and make plausible the new political option. This also seems to be the case with more recent movements such as Poujadism in France (Hoffman 1956) and the National Front in England (e.g., Walker 1977). What is clear, therefore, whatever the limitations and problems of studies such as Hamilton's, is that a functional as opposed to geographical explanation of new political movements such as Nazism is problematic and, in the light of "new" evidence, increasingly questionable.

PLACE AND POLICY OUTPUTS

In political sociology in recent years attention has shifted somewhat from an exclusive concern with the political process – elections, voting behavior, parties, interest groups, and political attitudes – to concern with the "outputs" of government: policies and services. Much of the research effort in this area has focused on subnational and local government. One reason for this is that it extends the "Who governs?" question into the ambit of "Who benefits?" (Clark 1973). This requires linking together attributes of localities with types and levels of policy outputs.

One assumption of these studies is that local governments can act more or less independently. As noted previously, this is an empirical question. Yet even in England, as Sharpe and Newton (1984) demonstrate quite clearly, the system of local government is sufficiently autonomous to justify an output study.

Many studies also assume that any particular local government could be located anywhere. The policy-making process is viewed as largely invariant (pluralist *or* elitist), and differences in outputs between local governments are seen as a function of demographic variables, "citizen-consumer" preferences (the Tiebout hypothesis) or budgeting procedures. Conventional approaches, therefore, fail to take place into account. Moreover, as Sharpe and Newton argue at great length, they also fail to explain.

By way of a substitute, Sharpe and Newton propose a perspective developed inductively: that the spending levels of cities on some services and in total reflect the positions these cities occupy in the national urban hierarchy, and their major economic activities and social structures. They strongly support the idea that cities are "holistic" entities rather than merely the sum of individual characteristics. They use the example of county towns and seaside resorts to make their point:

> Both city types were similar in terms of their population, in that they were generally wealthy, middle class, commercial and residential, and controlled by Conservative councils. Yet this similarity contrasts with a difference in spending patterns due to the fact that county types had a different economic base and function, and a different role in the system of cities, compared to seaside resorts (Sharpe & Newton 1984 : 150).

But other factors are also at work. As Sharpe and Newton's analysis of county policy outputs suggests, "cultural" differences can enter into the determination of public policy. They report that the Welsh counties formed a group distinct from the English ones, with much higher spending on local services. "There was, in other words, a Welsh effect which was special and unique to the principality, and which could not be reduced entirely to such things as class, politics, grants, sparsity, or the agricultural basis of the economy" (Sharpe & Newton 1984: 172). Clark (1975) has reported a similar finding from the United States, where the proportion of a city's population of Irish ancestry was more important in explaining expenditure levels than most other variables. Why? Because, it seems, in some American cities the Irish played a key role in creating strong (mainly Democratic) parties, obtaining patronage positions in city agencies, and enlisting the support of other immigrant, often also Catholic, groups in party primaries and bond referenda. The result of this has been increased public spending in these cities.

Finally, Sharpe and Newton (1984: 207) show that in general terms the various "objective" characteristics of a particular locality (age structure, population density, etc.) are "simply inert descriptive statistics . . . until they are given policy significance by the political system." The local political system abstracts some features from the local environment and gives them "political salience." Local politics is an active process of giving meaning, rather than a passive process of translating demands. Political parties are particularly important. They affect each other in terms of the local balance that exists between them, and select the policies which they will pursue in line with party ideology and the urge to get re-elected.

Somewhat ironically, then, the shift in focus from the political process to the analysis of outputs has, at least for Sharpe and Newton (1984) and some others (Ashford 1975, Peterson 1979, Saunders 1979, Hansen 1981, Aiken & Martinotti 1982, Parisi 1984, Hoggart 1986), led right back to a concern with the political process and, as an unintended consequence, with the uniqueness of places.

PLACE AND POLITICAL MOBILIZATION

A fundamental question in political sociology concerns the mobilization and politicization of groups of people on behalf of particular ideologies, movements, and political activities. Burnham (1982) has shown how electoral participation in the United States has waxed and waned over the past 100 years in response to the political choices made available and those denied by the party system. He argues that the recent trend towards a low salience of issues in political campaigns and declines in turnout are related. They reflect an increased political alienation: a conviction that party organizations are corrupted beyond redemption and out of reach of popular influence (Burnham 1982: 51). More specifically, however (p. 52), "a great many American voters, it would seem, are intelligent enough to perceive the deep contradiction which exists between the ideals of rhetorical democracy as preached in the school and on the stump, and the actual day-to-day reality as that reality intrudes on their own milieu." This is the "current crisis in American politics." In the absence of viable alternatives, the lack of perceived relevance of current American political parties to the interests and problems of those masses of Americans in inner cities and poor rural areas has depoliticized or taken out of the electorate close to half its potential population (pp. 191–2).

The study of participation and its converse – abstentionism, or nonparticipation – are important for what they say about political alienation, the possibility of disaffected voters shifting to new parties or surging into revived old ones, and the degree to which the continued integration of regimes is threatened by high levels of disaffection. They are also suggestive of the conditions under which political activity and struggle are produced. What is it that prompts members of some groups to act together in pursuit of common objectives when others do not? This may be *the* central question of political sociology.

The term "capacity" has been coined by several writers to refer to a group's ability to carry on political activities (e.g., Katznelson 1976, Wright 1978). Katznelson (1976) argues that a division of interests between workplace and residence has been central in the failure of socialism to develop in the United States. This illustrates an absence of sufficient "class capacity." In this instance, capacity involves the ability of a group to understand the conditions of its existence and to act accordingly.

How does capacity develop? As a first condition of group formation there must be a minimal recognition of common traits among potential members. The next essential element is within-group interaction. Sharp boundaries between groups encourage within-group interaction. Common life experiences and sustained interaction promote a common outlook or world view. This results in a capacity for collective action only if members recognize that their economic and political situation cannot be improved by individual action. Collective action itself is forthcoming only if group capacity is joined to organizational capacity. Organizations such as labor unions and churches educate their members about political options, the importance of concerted action, and the possibility of effecting change (Glantz 1958, Tilly 1978).

Certain settings or environments are much more conducive to the formation of a group outlook and collective action than others (Harris 1984a,b). It is a commonplace that certain industrial environments generate a strong sense of "working-class" identity with political behavior to match (Smith & Rodriguez 1974). Strong labor organizations and occupational– residential segregation have often been critical in such settings (Lockwood 1966). Such "production" cleavages, however, are reinforced or undermined by what Dunleavy (1979) has called "consumption" cleavages. Commitments to collective-service–public modes of providing such services as transportation and housing will act as reinforcements. Commitments to individualized-commodity–private modes of service provision will have an undermining effect.

But consumption can also be an explicit locus of group formation and collective action. A large literature has developed showing how access to public services has become a major focus of political activity in American cities (see Katznelson 1976, Rich 1981). In the presence of high levels of private provision, however, such as in American suburban areas, there may be a depoliticization of certain issues to the extent that the scope for political activity is considerably reduced except during times of "threat" (Newton 1978). Group formation in this type of setting, as it were, can lead to demobilization!

Production and consumption cleavages are then crucial elements in group formation and political mobilization. But there are two other aspects to political mobilization that remain at issue. One of these is the role of organizational capacity. The other is the types of political activity that mobilization produces.

Many studies have shown the importance of organizational capacity for politicization and increased levels of political activity (e.g., Tilly 1978, Benson 1980). Work in France on "peasant" politics and electoral abstentionism indicates that different places produce different degrees and orientations of organizational capacity (Lancelot 1968, Berger 1972). Berger (1972) has shown how the politicization of everyday life is a relatively recent phenomenon in rural Brittany. "Incomplete political integration" characterized the region until political issues were created that overlapped with those salient in the country at large. Only after this process was complete could national political parties mobilize local support. Integration is still incomplete except insofar as organizations exist to direct attention to national politics and away from more localized activities (Berger 1972:53-4). In Finistère local elites encouraged peasants in a corporative movement of syndicates and cooperatives without reference to the state or cities. In Côtes-du-Nord, however, the conservative elite managed the disruption of the established social order by mobilizing peasants to support particular political parties and partisan issues. Political mobilization presupposes this latter form of organizational capacity (see also Vedel 1962).

In a more general vein, Lancelot (1968) suggests that high participation in French elections is a function of the multiplicative influences in place of social homogeneity, political integration, work-place residence conformity and high levels of participation in trade unions or other intermediate groups. He summarizes as follows:

> The [social] categories that participate least in elections are also those which associate together least readily. The regions with the most voters are those where socialization into groups is the most widespread. In the department of Nord, for example, the high participation in elections cannot perhaps be explained without reference to an intense social participation (Lancelot 1968:226, my translation).

Political mobilization, therefore, is place-specific, reflecting the history of integration into a national political system (Berger), local organizational capacity, and other facets of group formation (Lancelot).

This can be viewed from the perspective of mobilizing organizations as well as from the perspective of places. Political parties, for instance, set agendas and identify social cleavages, and then organize to capture the support of the groups to whom they are appealing in the places in which those groups are concentrated. Congressional and parliamentary elections, in particular, are in most countries won locally, not nationally. As Johnston (1985b : 252) expresses it in connection with British elections:

> it is not the percentage of the national vote that matters but the number of constituencies ... A successful party in Britain is one that wins seats, and has an organization capable of delivering a plurality of votes in a majority of constituencies. It must have a strong, enduring local set of bases.

Of course, voting is not the only form of political activity. Why is it, for example, that political violence characterizes the political histories of some places but not others? Often this may have been the product of place-specific repression, or the absence of other alternatives such as electoral politics. However, it is clear that there can be settings in which political mobilization is more likely to produce violence or similar "dangerous" activities than others. Fitzpatrick (1978), in a study of revolutionary violence and nationalism in Ireland, demonstrates that it was the interaction between poorer and rural conditions and the incidence of branches of specific nationalist organizations, especially Sinn Fein, that explains the incidence of political violence in the period 1920–1. It seems that even the *forms* that political activity take are related to organizational capacity. And organizational capacity is place-specific. Political activities are therefore place-specific in origin also.

Conclusion

Considerable attention has been given in this chapter to arguing why place is of contemporary and not just historical importance. Attention has been directed to the range of research problems in political sociology that can be addressed from a place perspective. A wide range of literature has been reviewed to make this point. Some of this is from research *genres* in which a concept of place has been important. Much of it, however, is not.

It is clear from the examples pursued above that a place-based political sociology of sorts is already operative. But these fragments of a wider viewpoint have never been gathered together before, let alone placed in the context of a coherent and internally consistent theoretical framework such as structuration theory. Furthermore, the key research problems in political sociology have never been clearly associated with a place perspective. Yet many of these, such as the problem of political mobilization, changes in political behavior, the incidence of different types of political activity, the emergence of new political movements, aggregate patterns of political expression, and the political integration of current regimes, are hardy perennials. They are the central problems of political sociology. Many of them are raised in the later examination of Scottish and American politics from the place perspective.

It is quite remarkable how much empirical political sociology has suggested the importance of place without its authors making the connection. This has been the burden of the argument in this chapter. The next chapter is concerned with explaining why this, and the general lack of interest in place and politics in political sociology – and social science – should have occurred.

5 Devaluing place

It became widely accepted in political sociology in the 1950s and 1960s that, at least in the "advanced industrial democracies," if not elsewhere in the world, national patterns of political behavior had displaced ones that could be thought of as local or regional. In part this resulted from empirical research, such as that of Stokes (1967), which claimed to find that in both Britain and America there had been a progressive "nationalization" of political attitudes; he inferred from this that there was consequently a nationalization of electoral forces bringing about political attitudes. In part, however, it reflected a more deep-seated frame of mind concerning the significance of place in modern "national" societies as opposed to traditional "local" communities. Contemporary political sociology has inherited from its parent disciplines, political science and sociology, a set of biases against the possibility that place can be of any significance in modern societies. It may be of importance in traditional ones, but not ultimately since they too are fated to modernize. Place, it seems, has no present or future, only a past.

In this chapter, connections are drawn between the devaluation of place and larger intellectual features of social science in order to provide a perspective on why a geographical political sociology such as that proposed earlier has been impossible. The next chapter (Ch. 6) demonstrates how recent empirical work in political sociology supports the geographical interpretation; this chapter suggests why this has not usually happened.

The devaluation of place

There are two stages to the devaluation of place in contemporary social science and political sociology. The first involves confusing the geographical concept of place with the sociological concept of community. The second stage involves inferring from the supposed "eclipse of community" in modern society the waning significance of place.

THE CONFUSION OF PLACE AND COMMUNITY

Confusing place and community is widespread in political sociology. This reflects an underlying ambiguity in the language of community as it has developed in the social sciences. Much has been made of the language of class as a fundamental discourse in social theory. But at least as widespread today, and perhaps even more important in the 19th century, was the language of community. This language had an old heritage, but it was during the massive social changes of the 19th century that it came to bear two historically specific connotations that have persisted to the present day. Specifically, community was conceived as both a physical setting for social relations (place) and a morally valued way of life (Calhoun 1980:107). Thus (p. 106), "The language of community grew up as a demand for more moral relations among people as well as a descriptive category . . . Community was far more than a mere place or population."

Nisbet (1966:18) sees the moral usage of community as historically prior to that of social relations in a discrete geographical setting or place. He notes: "Community begins as a moral value; only gradually does the secularization of this concept become apparent in sociological thought in the nineteenth century." But even while sensitive to competing connotations Nisbet maintains the fusion of the "experiential quality of community" and the actual social relationships of place that has come to bedevil contemporary social science generally and political sociology more particularly. He goes on (p. 48) to argue strongly to the effect that:

> By community I mean something that goes far beyond mere local community. The word, as we find it in much nineteenth- and twentieth-century thought encompasses all forms of relationships which are characterized by a high degree of personal intimacy, emotional depth, moral commitment, social cohesion, and continuity in time. Community is founded on man conceived in his wholeness rather than in one or another of his roles, taken separately, that he may hold in a social order. It draws its psychological strength from levels of motivation deeper than those of mere volition or interest, and it achieves its fulfillment in a submergence of individual will that is not possible in unions of mere convenience or rational assent.

Of course, this viewpoint is not merely Nisbet's but draws on a long intellectual tradition. In Tönnies (1887) Gemeinschaft/Gesellschaft dichotomy, community had a similarly moral affirmation in the face of the demands of rational (and thus noncommunal and nonlocal associative) relationships such as social class. To Durkheim, the functions once performed by "common ideas and sentiments" in Gemeinschaft society were performed in large-scale industrial societies by completely new social institutions and relations. This transformation involved major changes in the nature of both morality and social solidarity. Durkheim's central thesis was that "the division of labor," by which he primarily meant occupational specialization, "is more and more filling the role that was once filled by the conscience commune; it is this that mainly holds together social aggregates of the more advanced type" (Durkheim 1933:173). The division of labor was "the sole process which enables the necessities of social cohesion to be reconciled with the principle of individuation" (Durkheim 1904:185). Weber (1925) took much the same view as Tönnies and Durkheim, seeing local community as based on subjective feeling in contrast to the rationality of national social relationships.

The founders of academic sociology in the United States were particularly interested in how the growth of cities affected the constitution of social relationships. Their "programmatic question," as Fischer (1975:67–8) has put it, reflected that of their European precursors: "How can the moral order of society be maintained and the integration of its members achieved within a highly differentiated and technological social structure?" It was "the problem of community in the New Age."

The typological approach of Tönnies seemed to offer an especially powerful insight into the "urban transformation" through which they were living (Faris 1967:43–8, Cahnman 1977, Shils 1970). Robert Park, for example, one of the leading figures in the sociology department at the University of Chicago and a founder of urban sociology, noted that though the diverse terms used to label the community–society dichotomy revealed its "unrefined nature": "the differences are not important. What is important is that these different men looking at the phenomenon from quite different points of view have all fallen upon the same distinction. That indicates at least that the distinction is a fundamental one" (quoted in Zimmerman 1938:81).

The reliance on the community–society dichotomy reached its peak in the urban theory of Wirth (1938) with its folk–urban polarity, and the modernization theory of Parsons (1951), with its traditional–modern dichotomy. Tönnies's idea that the modern world was the product of a transition from a place-based community to a placeless or national society was thus perpetuated in the dominant strands of American sociological theory.

In contemporary political sociology, for example, one finds a merging of place with community identical to that of earlier sociology. Hays (1967:154) provides an explicit example:

> One conceives of community as human participation in networks of primary, interpersonal relationships within a limited geographical context. At this level there is concern with the intimate and personal movement within a limited range of social contacts, and preoccupation with affairs arising from daily personal life. Knowledge is acquired and action carried out through personal experience and personal relationships.

But society is national, he continues, and involves "secondary contacts over wide geographical areas, considerable mobility and a high degree of ideological mobility." Society is therefore distinct and counter to community-in-place. When the "modernizing" forces of society overpower the "traditional" forces of community, place is overpowered too and continues to exist only as the location of nationally defined social activities. That modern *social* activities are or even could be defined in places has become a contradiction in terms.

Many writers involved in "empirical" political sociology are not as explicit as Hays about the meaning they ascribe to community. For example, in most of the research on the nationalization of the American (and other) electorates, community effects are simply viewed as local ones and vice versa (e.g., Stokes 1965, 1967; Katz 1973; Sundquist 1973, Converse 1972). Indeed, the terms "community" and "local" are often used interchangeably to describe district-level effects in turnout and vote choice. A decline in "community" effects signifies the declining significance of locality in national political life. So, though not explicit on the point of place versus community, the language used betrays the meaning.

However, students of political parties and electoral politics in general are often clearer (e.g., Lipset 1960, Lipset & Rokkan 1967, Rose & Urwin 1969). They frequently choose to use Parsons's scheme of "pattern variables" (Parsons et al. 1953) to distinguish place-based and communal from society-based and national patterns of politics. The former are characterized by particularistic, ascriptive, and collectivistic orientations to action, whereas the latter are generally associated with the values of universalism, achievement, and individualism. Clientelistic politics is seen as the "natural" outcome of place-based, communal and traditional values, with modern, democratic, Western, two-party mass politics seen as the "natural" outcome of society-based and national values (Lipset 1960). A very influential conceptualization of society and politics (Parsons et al. 1953, Parsons 1959) thus contains within it the confusion of place and community-as-a-way-of-life that has become characteristic of modern social science. This in turn has affected orthodox political sociology to the extent that many practitioners are now unable to see any distinction between them.

A few attempts have been made to distinguish place from community and argue for their noncomplementarity. For example, Webber (1964) has proposed spatial accessibility (ease of access) rather than the propinquity of place (coresidence) as the necessary condition for community (see also Bastide 1966). As Webber (1964 : 108–9) says:

> The idea of community . . . has been tied to the idea of place. Although other conditions are associated with the community – including "sense of belonging," a body of shared values, a system of social organization, and interdependency – spatial proximity continues to be considered a *necessary* condition.
> But it is now becoming apparent that it is accessibility rather than the propinquity aspect of "place" that is the necessary condition. As accessibility becomes further freed from propinquity, cohabitation of a territorial place – whether it is a neighborhood, a suburb, a metropolis, a

region, or a nation – is becoming less important to the maintenance of
social communities.

The problem with this is that it resolves the dualism of community in favour
of the way of life or solidarity element and at the expense of social relations
structured in place, despite tremendous evidence to the contrary. It also fails
to account for the fact that even in big cities in 'modern' societies, people act
together on occasion on the basis of common territory (Tilly 1973:212, Bell
& Newby 1971). Many communities still rest on propinquity.

At the crux of the confusion over place and community is the ambiguous
legacy of the term community. Rather than distinguishing its two connotations – a morally valued way of life and the constituting of social relations
in a discrete geographical setting – they usually have been conflated. In
particular, a specific set of social relations – those of a morally valued way of
life – have transcended the generic sense of community as place. Such an
emphasis has, as Calhoun (1978:369) reminds us, discounted "the importance of the social bonds and political mechanisms which hold communities
together and make them work." It has also reduced the possibility of seeing
society-in-place as an alternative to community-in-place. For orthodox
political sociology, and social science in general, place and society, and
consequently geography and sociology, have been antithetical

THE ECLIPSE OF COMMUNITY
The dichotomy of "community and society" has been a major framework
into which social scientists have set their discussions of human association
and social change. The theorists whose writings have provided the orientations and terms of discourse for much contemporary social science lived and
wrote during the last half of the 19th and early part of the 20th centuries. This
was a period of immense social change and perceived disorder in social
relations. The American and French Revolutions had set the stage with their
new doctrines of citizenship, equality and individual rights. The Industrial
Revolution was creating in Europe and America a new order of human
institutions and social classes. Populations had grown exponentially.
Nationalism had appeared as a principle of abstract solidarity. New states
emerged to change the political map of Europe. It was an era of economic
discontent and political revolt. Self-consciously, intellectual elites of the 19th
century were aware that what they had been brought up with was now passé;
that a new world of human relationships was coming into existence. It was
this experienced and perceived change in history that occupied the "social"
thinkers of the period (Gusfield 1975:3).

In the circumstances of the late 19th century it is understandable that a
major theme of many writers would be the *direction* of history. In particular,
a common belief in social evolution led many influential writers to view their
time as moving the world along a line of direction away from one point and

toward another. The movement involved was from one type of human association to another. They saw each as constituting a systematic arrangements of elements, and each as mutually exclusive from the other. The forms of the "modern" broke sharply with the dominant aspects of pre-19th-century life. The modern was both more individuated and individualistic in social organization than the "traditional." Bonds of group loyalty and affective ties had given way to the rationalistic ties of utilitarian interests and uniform law (Gusfield 1975:4–5).

Where writers differed was in their reaction to the movement from traditional to modern, from community to society. For some the "loss" of community was a lamentable feature of the period. For traditionalists, most especially Comte perhaps, the new order of economic logic, political interest groups, mass electorates, and exchange-based market relationships represented the destruction of a stable social environment and traditional social hierarchies essential, in their view, to human organization. Others, however, viewed the change in a more positive light. Writers as different in other respects as Spencer and Marx saw community as coercive, limiting, or idiotic. In its place they saw the possibility of human equality and economic affluence as society overcame the limits and constraints intrinsic to community.

But whether nostalgic or ecstatic, these theorists shared a belief in the 'eclipse' of community in 19th-century Europe and America. Some of them. however, were careful in how far they would take the point. Weber (1949), for example, was quite specific in arguing that ideal-types such as community or society are not averages or descriptions, but models, useful in understanding reality but not to be mistaken for it. In current usage the concepts of community and society are opposites in an almost zero-sum form. This has involved reifying what started out as ideal-types or analytical terms into empirical descriptions (Bender 1978:32–40). There are several areas of contemporary social science in which this usage of the community–society distinction is most apparent. One is the study of "development" in "newly emerging" but economically "underdeveloped" countries. The other is political sociology.

The area of development studies, which includes work in various disciplines, has made considerable use of the concepts of tradition and modernity in attempts to understand and assess contemporary change. One particularly influential writer, Redfield (1930, 1955), described a contrast between ideal-type "folk" communities and "urban" societies. Although Redfield was careful to note the empirical coexistence of the two forms, his and similar work was taken up by a later generation of writers motivated by an urge to make their efforts relevant to the post-World War II world of new nations and the decline of European colonialism and less careful about using ideal-types (Rostow 1960, Almond & Coleman 1960, Lerner 1958). The problem was, how could new nations overcome poverty and political instability to achieve economic growth and political stability? Equipped with the evolutionary perspective permeating modern social science, they wrote of modernization, educational development, and political development. Each of these processes involves as a central moment the total movement away from community to society. The pitting of one against the other, however, overlooks the mixtures or blends characteristic of actual settings. Moreover, it becomes an ideology of "antitraditionalism, denying the necessary and usable ways in which the past serves as support, especially in the sphere of values and political legitimation, to the present and the future" (Gusfield 1967 : 362).

The "nation building" focus of the development theorists is also an important element in orthodox political sociology. Figures such as Lipset and Almond are indeed important contributors to both literatures. Lipset's (1960) widely quoted essay on "Economic development and democracy" is a classic in the genre of modernization theory as well as an important background chapter of *Political Man*, a central statement of orthodox political sociology. Elsewhere and more recently Lipset (1981) tries to explain "backlash politics" or the rebirth of protest politics in Europe and the United States as a "revolt against modernity."

Whereas many 19th-century theorists bemoaned the passage of the past in the form of community, the modernization theorists find this a welcome transformation. It is now prescribed as the means to overcome backwardness, deprivation, and penury. The evolutionary sequence suggests that the communal social system *must* retreat and give way before the impetus of the modern if modernity is to be achieved. In its present use, therefore, "community" is again an ideological counterpoint to existing societies. To writers such as Comte or Tönnies, community signified a world that was lost, whose disappearance highlighted the tragedy of modernity. In modernization theories it is a counter once more, but now as a world, if not yet eclipsed, to be eradicated in the interest of "modern society."

A number of writers have pointed up the centrality of discourse based around the community–society dichotomy in modern social science (e.g., Gusfield 1975, Tilly 1973). But they have also suggested its essential limitations. Tilly (1973), for example, notes that communities, by which he means "any durable local population most of whose members belong to households based in the locality" (Tilly 1973 : 212), have not declined absolutely. Rather, as *places*, these have persisted in importance. But there has been a *relative* decline of such communities as bases of collective action: "local ties have diminished little or not at all, extralocal ties have increased" (Tilly 1973 : 236). Society is experienced in place and community remains a *part*, if a diminished part, of this. Tilly concludes:

One can notice the rising proportion of extralocal contacts and regret it. One can compare present conditions of solidarity with an ideal integrated folk community and find the present wanting. But on the basis of present evidence one cannot claim that urbanization [or modernization] produced an absolute decline in community solidarity.

Of particular importance, and beginning in the 16th century, the formation of national states, the growth of international economic markets and industrial concentration have clearly led to massive social change (Tilly 1979). In consequence, places have undergone *fundamental* change in the relative importance of local as compared to extralocal ties.

But Gusfield (1975) stresses the continued significance of "the communal impulse" in the face of dramatic social change. Unlike Tilly, who argues for the persistence of community-as-place and the new importance of society-in-place, Gusfield sees evidence for the "interactive character of much communal and societal organization"; the communal is now interwoven with the *national*–societal:

> Granted that there are economic interests which divide the population, the relation of these to communal bonds of both obligation and opportunity are of central importance to the mobilization of conflict and to further maintenance and change within a system of stratification. *The mix of the two communal and societal types is of more importance than their differences* (p. 68, Gusfield's emphasis).

Thus, rather than being "eclipsed," community survives and is diffused within society in the form of the specific social relations classic theorists associated with the morally valued way of life (community) they saw in decline (Bernard 1973).

Earlier, reference was made to a landmark study in political sociology by Stokes (1967). Stokes argued that the American electorate has become increasingly nationalized over the past 100 years. His major evidence was a steady, century-long decline in the impact of constituency-level forces on turnout in congressional elections, and a more recent and smaller decline in the impact of district forces on partisan voting (Stokes 1967:192–8). Others using less sophisticated methodologies than Stokes, such as Schattschneider (1960:78–96), Sundquist (1973:332–40) and Sorauf (1980:51), have come to similar conclusions. Recent research, however, suggests that these conclusions are unwarranted (Mann 1978, Pomper 1980:87–91, Claggett et al. 1984), even though they have become part of the conventional wisdom of American political sociology (see, for example, Converse 1972:268, Niemi & Weisberg 1976:447–8, Polsby 1981:29). Using Stokes's methodology, Claggett et al. fail to find any increase in the nationalization of partisan voting over the past 100 years, either as a convergence in levels of partisan support or in the uniformity of response. Stokes's conclusions on turnout are less problematic.

Why should there be such a disparity in findings between studies using the same data and the same methodology? Claggett *et al.* (1984:78–9) suggest that one element in a complete explanation is the partial nature of Stokes's analysis:

> Stokes did present data on the national effect for turnout and, although not as smooth as the constituency effect, overall an increasing trend can be seen (Stokes, 1967, p. 194, Figure 3). However, no corresponding data were presented for the national effect for partisan voting, although a footnote promised such data in a future article (Stokes, 1967, p. 196).
>
> This national effect for partisan voting in congressional elections can, however, be calculated very simply with the use of Stokes's method . . . It is clear that there is no substantial increase in the national component over the years from 1872 to 1960.

Elsewhere in their paper Claggett *et al.* (1984:77) suggest that Stokes had found what he thought was a trend in partisan voting and felt no need to examine it further, perhaps because it and the trend in turnout seemed "so consistent with the decline of parochialism and localism under the onrush of modernization". In their conclusion (p. 90) they elaborate on this:

> The plausibility of the nationalization argument rests on television, along with rapid transportation and other forms of communication, creating a national political environment and bringing the same effects and issues to the attention of people (and voters) everywhere – in short providing a common set of national stimuli. But then the argument makes something of a leap: not only does television provoke a common set of national stimuli, it also provokes a common set of responses.
>
> . . . to the extent that the salient political cleavages in a society follow, rather than cut across geographic divisions, a national political environment and national media focusing attention on the same issues everywhere may serve to accomplish the opposite of nationalization, that is to stimulate dissimilar behavior on the part of geographic units.

To a considerable degree, therefore, the empirical basis to nationalization rests upon an a priori belief in the eclipse of community and the declining significance of place. Having confused place and community, place has then been defined as characteristic of communal association. The supposed eclipse of community has in turn led to the eclipse of place. Thus has social science in general and political sociology in particular effectively and systematically devalued place as a concept relevant to our time. Its association in the academic mind with parochialism and localism has become so deep-rooted that the idea of place as the structuring or mediating context for social relations seems strange and out of temper with the national focus of most contemporary social science.

Explaining the devaluation of place

Why has the intellectual devaluation of place occurred? Hints at an explanation have appeared earlier in this chapter. But it is an important and useful task to systematically sort out a "chain of explanation" from intellectual to social–historical sources of devaluation in orthodox social science. "Antithetical" social science, in particular Marxist political economy, is then identified as a further but distinctive source of place devaluation.

The argument with respect to orthodox social science proceeds as follows: first, the most immediate intellectual source of devaluation is seen as the focus on the evolutionary sequence of transition from community to society. Secondly, an important correlate of the evolutionary perspective is identified. This is that the transition is natural, lawful, and universal. Thirdly, the key historical period for the intellectual devaluation of place is located in the 19th century, an important period in the growth of nationalism as a place-transcending ideology. Fourthly, it is argued that the extension of placeless social science in the form of modernization theory occurred in the historical context of the Cold War and the "struggle for hearts and minds" between the West and "World Communism" in the Third World of developing nations.

Attention then turns to antithetical social science in the form of Marxist political economy. The treatment of place in this literature reflects a focus on the practical process under capitalist relations of production of "freeing" people from places, and rationalizing such activities in terms of economic efficiency and market rationality. At the "deepest" level, so to speak, the intellectual devaluation of place in Marxist political economy is a product of an analysis that *accepts* the logic of liberal capitalism and its associated features of individualistic atomism and commodification of people and places.

THE EVOLUTIONARY SEQUENCE
As previously argued, at the root of the intellectual devaluation of place in orthodox social science is the oppositional dichotomy of community and society and its image of total temporal discontinuity between two completely different forms of human association. In this usage, the concepts of community and society embody a specific theory of social change, that of the evolutionary transformation of life *from* community *to* society. When community and society are seen as opposites, coexistence is transitory and life *must* move from one end of the continuum to the other. As society displaces community in this process, place loses its significance since it is closely intertwined with community.

In bare essentials, therefore, the evolutionary sequence of movement from community to society provides the intellectual backbone to the devaluation of place in orthodox social science. But though a necessary element in a complete explanation, in itself this is not sufficient. In particular, there is no reference to why this approach should have become so compelling, and why it did so when it did.

THE ANALOGY TO NATURE

An important feature of late 19th-century social thought was its use of ideas from the natural sciences of the day. Evolutionary ideas from biology, particularly those of Lamarck and Darwin, were especially attractive to social thinkers. Motivations for drawing on these fields were mixed, but one powerful encouragement was that natural explanations – ones drawing by analogy or homology from the natural sciences – would be free of the tinge of religion, ideology, and "free opinion" that had previously characterized social thought. Such explanations would also be universal, and thus congruent with the tenets of the empiricist philosophy predominant amongst "scientists" and philosophers at that time.

Use of the term "natural" is potentially confusing in the context of a discussion of community and society. Proponents of society generally maintain the usage of the Physiocrats, and have in mind unconstrained and self-regulated social order such as that found in nature. Proponents of community, usually critics of the status quo and thus easily portrayed as an ideological lot, use natural more in the sense of simple, primitive, or close *to* nature. In both cases, however, the former "scientific" and the latter "ideological," nature is invoked as the appropriate standard and grounding for explanation.

One can trace the scientific analogy to nature amongst social thinkers back at least to the 17th century. Many Enlightenment thinkers, perhaps especially the Physiocrats, aspired to explanation of social facts in terms of natural processes or by analogy to natural processes (Weulersse 1918, Fox-Genovese 1976). Viner (1960:59) has noted of the Physiocrats that they

> arrived at their laissez faire doctrine by way of a curious blend of the myth of a beneficent physical order of nature, of Hobbesism, of Cumberland's and Cartesian rationalism, and of some fresh and important economic analysis of the coordinating, harmonizing, and organizing function of free competition. There was a providential harmonious and self-operating *physical* order of nature, which, under appropriate social organization and sound intellectual perception, could be matched in its providential character, in its automatism, and its beneficence, in the *social* order of nature.

The imitation of scientific discourse encouraged a strong tendency towards objectification. Instead of sacred symbols and rituals as the expression of collective existence, the essence of society is seen to be production and subsistence. Society is reconstituted in terms of functional relationships between *individual* men and things. Indeed men become indistinguishable from things. At the same time, conceptions of natural law exclude the possibility of exception or idiosyncrasy. Natural law is associated with recurrent regularities, uniformities among phenomena, and classifications which omit singularities or differences (Crocker 1959, Wade 1977, Meek 1976).

In this intellectual milieu, social science found its first distinctive voice not as a new theory of society or politics but as a discourse about economics. Although economic ideas have roots that can be traced back to Aristotle's discussion of *oikos*, one source of modern economics lies with the Physiocrats. Their ideas were later absorbed into a number of different theories. Physiocracy, *physis* (nature) + *kratos* (rule), can be translated as nature's regime or as that form which embodies nature's rule. The Physiocrats discovered the economy to be a submerged order, suffocating under the burden of government and mercantilism, diverted from its natural course. The economy was thus an immanent order, ready to emerge and function according to laws inherent in the nature of things but stifled by politics (Meek 1976).

In asserting that free economic activity was "natural," the Physiocrats transferred to the plane of economics the cosmological understanding of a universe in which phenomena, social as well as physical, obeyed the laws of their nature, a nature perhaps initially implanted by God but now operating as a natural necessity without need of an "external" power to cause or correct it (Meek 1976, Livingstone 1984). The freeing of the economy, therefore, became equivalent to the liberation of nature.

This is not to say that 19th-century theorists were of the same mind as the Physiocrats or one another when it came to either their views of laissez faire or nature, or how to cast their analogies (compare Spencer and Marx, for example). However, many social theorists shared the underlying belief that the analogy to nature was essential to *scientific* explanation. Marx, for example, counterposed his scientific socialism to the socialism of those he characterized as Utopians (e.g., Proudhon), largely on the presumed grounding of his argument in the natural order of things.

It is in the intellectual context of the naturalness and lawfulness of social life that the evolutionary perspective on the transition from community to society must be seen – not only because the movement itself can be seen as natural and lawful, but more especially because, at least in the hands of some theorists, the world of society is also natural and lawful when compared to the world of community. This implies, of course, that the analogy to nature is *always* social or political rather than scientific. In Gusfield's (1975) terminology, concepts such as society or community are "utopias" as well as ideal-types. The active pursuit of society, however that concept is rendered, is justified as one with nature. How can nature be doubted?

THE CONTEXT OF NATIONALISM

In the 19th century, the analogy to nature favored by many social theorists had to coexist with the growth of nationalism as a powerful political ideology. This led to an interesting convergence between the two. Society, rather than remaining an abstraction or ideal-type, became coterminous with the boundaries of national states . . . and naturally so. Herder (in Barnard 1969) provided a clear connection ("a nation is as natural a plant as a family, only with more branches") and many social theorists followed suit. A principle of what Smith (1979 : 191) calls "methodological nationalism" came to prevail.

The three disciplines that were the major winners in the institutionalization of the study of social life – sociology, political science, and economics – each developed a subservience to the state. Indeed, each had at its origins the practical interests of the state – respectively, in social control, state management, and the national accumulation of wealth (Nisbet 1966 : 17, Skinner 1978 : 350, Letwin 1963 : 217, Deans 1978 : 203, Robinson 1962 : 24–5). At their roots, therefore, they were national in focus.

Many of the most influential thinkers of the 19th century were methodological nationalists. Whatever their other differences, Marx, Durkheim, and Weber all accepted state boundaries as coextensive with those of the "societies" or "economies" they were interested in studying. More particularly, to thinkers such as Durkheim the modern state was both creator and guarantor of the individual's *natural* rights against the claims of local, domestic, ecclesiastical, occupational, and other secondary groups. The state could thus function both as an "enforcer" of natural order as well as a "container," through the territorial definition it provided, of empirical observations about social and economic processes (Cohen 1982, Woolf 1984).

Other disciplines such as ethnography, anthropology, and human geography served the state through the knowledge they provided about potential and actual "colonial societies" (Ranger 1976). These were ones in which community (and place) were still important. At "home," however, these disciplines, especially human geography, provided a rationale for territorial delimitation of the state (so-called "natural boundaries") and, notoriously, in the form of *geopolitik*, a rationale for state expansionism (see Kasperson & Minghi 1969, Parker 1982).

By and large, therefore, the social sciences have been oriented towards the national state as a "natural" unit upon which to build their claims to generalization. This presupposed the diminished importance of community on a local scale, and the social significance of place along with it. Even attempts at "reviving" community, perhaps most visibly in the hands of national dictators such as Mussolini and Hitler, but also more generally, have tended to operate solely at the national scale (Mosse 1975, Hobsbawn & Ranger 1983). Reference to local settings or to global processes was largely closed off by the nationalizing of social science and its subservience to the national state.

THE "THREE WORLDS" OF SOCIAL SCIENCE
In his study of early 19th-century America, Tocqueville (1945 : 14–18) argued that "democratic" nations have a greater capacity for general ideas than "aristocratic" ones. In abandoning the categories of the *ancien régime*,

people were left to invent new categories with which to explain and give meaning to their experience. As he put it,

> When I repudiate the traditions of rank, professions and birth, when I escape from the authority of example to seek out, by the single effort of my reason, the path to be followed, I am inclined to derive the motives of my opinions from *human nature* itself, and this leads me necessarily, and almost unconsciously, to adopt a great number of very *general notions* [my emphasis].

Pletsch (1981) has suggested that Tocqueville's reference to the propensity for general ideas provides an important perspective on the proclivity of social scientists to debilitate public discourse by imposing simple typologies on a complex social reality. One simple typology he selects for investigation is the taxonomic idea of three worlds that many of us use to divide up the world and its inhabitants. This "instance of primitive classification" is used by Pletsch to illustrate an intersection between politics and social science in the 1950s and 1960s that gave a boost to modernization theory and thus, from the present viewpoint, contributed to the intellectual devaluation of place.

In the aftermath of the Second World War, Pletsch (1981 : 569) argues,

> a great variety of social scientists and even journalists from several different nations, diverse ideological perspectives, and academic disciplines suddenly found the idea of a third world useful for organizing their thinking about the international order that had emerged from the settlements and unsettlements of World War II. And understanding the new order, they could place the significance of their own particular research in the grand enterprise of understanding social phenomena in general.

He traces the origins of the term "Third World" to a pair of binary distinctions. First, the world is divided into "traditional" and "modern" parts. Second, the modern part is divided into "communist" and "free" portions. These terms derive their meaning from their mutual opposition rather than from any relationship to real-world phenomena. The traditional world is the Third World, for example, whatever the situation "on the ground." Thus,

> The Third World is the world of tradition, culture, religion, irrationality, underdevelopment, over-population, political chaos, and so on. The second world is modern, technologically sophisticated, rational to a degree, but authoritarian (or totalitarian) and repressive, and ultimately inefficient and impoverished by contamination with an ideologically motivated socialist elite. The first world is purely modern, a haven of science and utilitarian decision making, technological, efficient,

democratic, free – in short, a natural society unfettered by religion or ideology (Pletsch 1981 : 574)

The distinctions are not only ideal-types, they also imply a historical relationship between the three categories they generate. Traditional societies are all destined to be modern ones somehow and to some degree. That has been the trajectory of history. That is how the First World has emerged. The Second World is an unfortunate perversion of "natural" modernization. The Third World is presently the zone of competition between these two models of modernity.

To Pletsch, the cold war context of the 1950s and 1960s is the vital backdrop to the emergence of the social science schema of three worlds. Not only because of the boost it gave to modernization theory and the conventional teleology of "from community to society." More especially, that even

> our strange taxonomy of social sciences – the belief that there exist such distinguishable things as politics, economics, and society, and that there should be a separate group of social scientists to study each – has been subordinated in the last three decades to the idea of the three worlds (Pletsch 1981 : 567).

As a consequence a new academic division of labor emerged. Thus,

> one clan of social scientists is set apart to study the pristine societies of the third world – anthropologists. Other clans – economists, sociologists, and political scientists – study the third world only insofar as the process of modernization has begun. The true province of these latter social sciences is the modern world, especially the natural societies of the West. But again, subclans of each of these sciences of the modern world are specially outfitted to make forays into the ideological regions of the second world (Pletsch 1981 : 579).

This has served to reinforce the view of Western societies as the ones most amenable to nomothetic social science, and consequently the ones in which reference to place is particularly inappropriate.

Tocqueville, then, usually read only as philosopher of early-19th-century American localism, provides a useful insight into why "modern" American social scientists, through their proclivity for "general ideas," have been attracted to a view of the world in which place is unimportant if not invisible. In the natural, democratic and economically free nations of the West, "the Free World," the concept of place *can have no role whatsoever*. They are now "beyond place," so to speak. The nation and natural law rule it out.

"THE LOGIC OF LIBERAL CAPITALISM" AND PLACE

So far, the intellectual devaluation of place has been accounted for in terms of a particular set of intellectual predispositions and contributory historical–

political factors. This certainly fits the bill for orthodox social science. But why is it that antithetical social science, especially Marxist political economy, usually opposed to the major tenets of the orthodoxy, has arrived at a similar devaluation of place? Of course, this viewpoint shares the evolutionism and naturalism of the orthodoxy. But thereafter they part company.

The major focus of Marxist political economy is on the logic of liberal capitalism. This logic postulates that most activities and goods are of no or limited value if they cannot be used in exchange. By distinguishing exchange value from use value, the value in exchange from the value in use of an activity or good, it is possible to argue that "pure exchange value" through the medium of money provides the basis for the practical devaluation of place. This is so because the process of circulation of commodities only reaches an intensive level if exchange value is detached from goods and exists as a commodity in its own right – that is, money. Money makes possible the circulation of exchange values across large distances and, in turn "makes possible the disembedding of social relationships from communities of high presence-availability" [local ties] (Giddens 1981 : 115).

Giddens goes on to provide a useful analysis of what he calls the "money–commodity–money" relation and its implications for place (pp. 109–28). He draws on Marx's *Grundrisse* and Simmel's "The philosophy of money", however, to argue that all commodities have a *double existence*, as "natural product" and as exchange value.

First of all, there is a practical basis to the devaluation of place. This can be seen in the "detachment" of people from places in the form of the commodification of land and labor (Pred 1983). Giddens (1981 : 118) argues:

> Labour-power, as a commodified form, relates to labour in the traditional sense much as money relates to the use-values of goods for which it is exchanged. As a commodity, labour has a similar double existence to other commodities, as on the one hand the expenditure of human skills and abilities, and on the other a "cost" to capital, defined in terms of its value in exchange.

An important consequence of this is the conceptualization of people as a "factor of production," labor, on a par in the economic process to land and capital. People and places then become commodities; people as labor and places as locations.

However, and this is the main point, the "double existence" of labor-power as a commodity provides an important barrier to the "inexorable" logic of capitalism as the motor of modern history in two ways (Giddens 1981 : 120). First, labor-power is the only commodity which itself produces value. Secondly, the "other" aspect of labor-power (its noncommodity form) "is not merely the use-value of a material good but a living human being with needs, feelings and aspirations. Labour-power is a commodity like any other – but resists being treated as a commodity like any other." Consequently, "labor-power consists of the *concrete activities* of human beings, working in *definite industrial settings*, who resist being treated as on a par with the material commodities which they produce" [my emphasis].

Place has been transformed. It is not *just* a question of simple solidarity by local reference groups in the face of "relations of absence," or "outside" influence, although this continues. Rather, to quote Giddens (1981:121) again,

> The vast extension of time–space mediations made structurally possible by the prevalence of money capital, by the commodification of labour and by the transformability of one into the other, undercuts the segregated and autonomous character of the local community of producers. Unlike the situation in most contexts in class-divided [but non-capitalist] societies, in capitalism class struggle is built into the very constitution of work and the labour setting. In the context of the productive organisation, whatever sway the wage-worker gains over the circumstances of labour is achieved primarily through attempts at 'defensive control' of the work-place.

Class relations are thus structured through the *place-based* exigencies of commodification and resistance.

Moreover, as Massey (1984:117) has argued in considerable detail, "geographical variation" in economic activities "is profound and persistent." In particular, "spatially-differentiated patterns of production are one of the bases of geographical variation in social structure and class relations." But the "uniqueness of places" is not simply an "outcome." Many industries locate different parts of their production processes in different places to take advantage of the historically sedimented and peculiar attributes of different places. One of these might be rates of worker unionization, another could be the availability of skilled workers. Rather than annihilating space, therefore, the "logic of liberal capitalism" has led to a permanent but dynamic process of uneven development.

Commodification of land and housing reinforces this process. Rather than obliterating place, commodification of housing has stimulated an intense concern by home owners in particular over *local* events that threaten established use and exchange values (Agnew 1981c). More generally, all people are trapped in a struggle over the appropriation of all values, as geographical patterns of external costs and benefits change in response to shifts in investment. Cox (1985) refers to this as the "politicization of location" from which has emerged a "spatial politics" based upon competition between places for new investment and a positive balance of external costs and benefits (see also Cox 1978, Logan 1978).

Thus, rather than disappearing in the face of the logic of liberal capitalism, the *nature* of place has changed. The *balance* between "relations of presence" (local ties) and "relations of absence" (extralocal ties) has moved in the direction of the latter. But this has not led to the demise of place. Capitalism, while transforming society, has created a *new* structuring role for place (cf. Tilly 1978, 1979).

In *practice*, therefore, the logic of liberal capitalism has been both limited and resisted, and has also created new pressures for place. In classic Marxist political economy this has not been recognized. Rather, as Appleby (1978:21) has pointed out in her discussion of the origins of economic thought: "Abstractions describing commercial transactions had become more real in men's discourse than the tactile and concrete context in which they happened." The lack of attention to the "double existence" of commodities has been particularly misleading. It has led directly to the view characteristic of orthodox economists, but shared by most theorists of "modern," "industrial," or "capitalist" societies, that concepts such as place are vestigial, useful perhaps for examining "past" societies but irrelevant to present or future ones. The focus on commodification has also missed the point that *dominant* social orders do not preclude the survival or creation of alternate ones in which other processes mediated in place are operative (Williams 1980).

Conclusion

The intellectual devaluation of place characteristic of orthodox social science in general and political sociology in particular has been traced in this chapter to a combination of influences: sociological, political, and intellectual. A static view of place, in particular its close association with community, has led over the previous century to an intellectual stillbirth. This reflects in part the enshrinement of the community–society metaphor as a major model of social change in orthodox social science, its "naturalization" as a scientific explanation, and "nationalization" as a political explanation. The growth of modernization theory as a weapon in the ideological combat of the Cold War added another nail to place's intellectual coffin.

A powerful impetus to the devaluation of place has also come from "antithetical" social science, in the form of Marxist political economy. In that literature there has been a tendency to absolutize the power of commodification and thus to see little scope for place in contemporary society, or by extension in contemporary social science. The arguments of Giddens (1981) and others suggest, however, that though changed and changing in terms of the social relations it structures, place has a continuing relevance for understanding the workings of modern, capitalist society. Both orthodox and antithetical social science, therefore, have disposed of place peremptorily. It is past time for its rehabilitation.

6 Discovering place

Since the late 1960s, confidence in the nationalization thesis has been shaken. One cause of this has been the emergence and success of regionalist and separatist movements and parties in many parts of Western Europe and in Canada. Another has been the persistence or strengthening of place-specific patterns of voting behavior in elections and referenda (e.g., Madgwick & Rose 1982, Miller 1984). Perhaps the most important, however, has been the calling into question of existing states as the sole legitimate focal point for the explanation of popular political behavior (Rokkan & Urwin 1982). But this has not led to the sort of theoretical framework outlined above. The usual response has been to "patch up" or add a new "factor" or two to established modes of theorizing. In the most satisfactory response, a territorial (or "sense of place") dimension has been introduced as a cleavage additional to those of class, religion, and so forth, but used in the same way.

This chapter is intended to fulfill three objectives: first, to describe the nationalization thesis in more detail than was possible in Chapter 5; secondly, to review evidence from studies in Europe and North America *against* nationalization; and, thirdly, to describe and evaluate various approaches to explaining geographical variation in political behavior.

The nationalization thesis

A widely accepted if not dominant *premise* of modern political sociology is that political alignments have crystallized around *national* social cleavages to produce *national* patterns of political mobilization and partisanship. In the late 1960s, Lipset and Rokkan declared that Western European party alignments had stabilized with the advent of mass suffrage and remained static for the next fifty years (Lipset & Rokkan 1967). Rose and Urwin's (1969) study of persistence and change in nineteen liberal democracies from 1945 to 1968 presented the same picture of enduring alignments and stable social bases in party voting. In this model of political modernization, geographical political cleavages are given either primordial status or viewed as following more fundamental social cleavages (some areas have more working-class people, others have more middle-class people, and so on). If primordial, they are "territorial." They date from the emergence of the modern state. But as the state survives and prospers, as industrialization and urbanization occur, citizens are mobilized into a national political "community." Allegiances are transfered from the local to the national community, and the nation-state becomes the preeminent political unit. Once this has occurred, "conflict is no longer between constituent territorial units of the nation, but between different conceptions of the constitution and organization of the national party" (Lipset & Rokkan 1967 : 23).

According to the advocates of this model of political modernization, it is exemplified by the United States. A 1906 speech by Elihu Root (1916 : 366–7), the US secretary of state, provides one of the clearest statements of the nationalization thesis from an American perspective:

> Our whole life has swung away from the old state centers and is crystallizing about national centers ... The people move in great throngs to and from state to state and across states; the important news of each community is read at every breakfast table throughout the country; the interchange of thought and sentiment is universal; in the wide range of daily life and activity and interest the old lines between the states and the old barriers which kept the states as separate communities are completely lost from sight. The growth of national habits in the daily life of a homogeneous people keeps pace with the growth of national sentiment.

The redistribution of population and income ,has reduced social and economic variation among the states (Haider 1974 : 24, Hofferbert 1971). Geographical mobility has, according to McWilliams (1972 : 32) "weakened the attachments of Americans to home and place; for increasing numbers ... a state or city is only a location where one happens to live at the moment." National types of economic organization, consumption, and leisure activities have come to predominate. Urbanization and economic change have brought even the South, the bastion of territorial distinctiveness, into the national mainstream (McKinney & Bourque 1971).

Accompanying the nationalization of social life has been a corresponding nationalization of American politics (Campbell et al. 1966 : 213, Hofferbert & Sharkansky 1971, Haider 1974). Irish and Prothro (1968 : 145) observe that Americans "talk about *national interests, national purposes, national goals* even in areas long-regarded as local concerns, such as health, housing, education, farm prices, labor standards, and civil rights." Territorial, or any other real, as opposed to epiphenomenal, geographical, cleavages have disappeared from political life.

But this correlation of political modernization and nationalization has also been a standard feature of writing about political behavior in the countries of Western Europe. The central theme of many textbooks on British politics is national homogeneity. To Blondel (1963 : 26), for example, "Britain is essentially a homogeneous nation in which the major distinctions are not based on geography, but on social and economic conditions ... in Britain, national class differences are the main divisions of society." Even local conditions are much the same everywhere. As Pulzer (1967 : 43) puts it, "the conditions and preoccupations of life are much the same in Bristol and Bradford" (see also Nossiter 1970). That this is not just a curiosity of the 1950s and 1960s is illustrated in a recent book by Bogdanor (1983) and a recent paper of McAllister (1987).

Many Anglo-Saxon and French commentators on French politics have also adopted the political modernization model. Duverger (1973), for example, argues that "archaic forms" of society that preexist the national state can survive for a while alongside it but ultimately disappear. The national "culture" eventually defines the terms and circumstances of political expression. In a widely hailed book on the "modernization" of rural France, Weber (1976) provides some Anglo-Saxon input. In the period 1870–1914, Weber argues, peasants became Frenchmen. From diversity and particularity, not to say barbarity, emerged homogeneity and civilization. This occurred because as France grew more prosperous, roads, railways, markets, schools, national newspapers, and military conscription penetrated the countryside. These rationalizing and nationalizing institutions undermined rural particularism, opened the hinterlands to new ideas, goods, and practices, and tied the countryside into the national culture and social life.

Some commentators on Italian politics have argued for a "territorial homogenization" in levels of support for the major Italian political parties (Mannheimer 1980, Brusa 1983). This view became especially popular between 1972 and 1979, though it has receded somewhat in recent years (Brusa 1984b, Pavsič 1985). The major thesis was that the three largest parties, the Christian Democrats (DC), the Communists (PCI), and the Socialists (PSI) were developing from their previously regional bases into national parties.

Although the definition of nationalization is rarely precise, two different aspects of it are apparent. The first of these is that nationalization is the convergence in levels of partisan support across subareas (constituencies, regions) producing a more *uniform* electorate geographically. This may not involve homogenization of the population, but does suggest increasing similarity in the mix of political sentiment across a nation. For the United States, Schattschneider (1960:87) adopted this viewpoint. He wrote

> elections since 1932 *have substituted a national political alignment for an extreme sectional alignment everywhere in the country except in the South*. Graphically, the nationalization of American politics can be seen in *the flattening of the curve showing the percentage distribution of the major party vote outside the South* (emphasis in original).

The second aspect, and perhaps the most visible one among political sociologists, sees nationalization as a uniform response to political forces. The universality of political trends – shifts or swings in turnout and votes from one party to another or others that are increasingly uniform across subareas – is taken as the measure of nationalization. Probably the most influential work in giving empirical support to the second definition of nationalization is that done by Stokes (1965, 1967) on American electoral behavior. He measured the relative magnitudes of national, state, and constituency effects on turnout and vote choice in congressional elections. The national effect measures the variance accounted for by the uniform movement of congressional districts across a set of congressional elections. The state effect measures the variance explained by the uniform movement of congressional districts within each state in those elections. The third effect measures the residual variance, which is presumably related to district-level influences.

In his 1965 article, Stokes reported that national effects accounted for far more of the total variance in turnout during the 1950s than either state or district effects. But district effects accounted for the largest proportion of variance in the division of the vote between parties, with the national effect a close second (Stokes 1965 : 75–8). In the 1967 article, Stokes concluded that the American political system had become increasingly nationalized since the Civil War. His evidence for this is a steady decline in the relative importance of district effects on turnout and a decline, if smaller and more recent, in their significance for the partisan vote (Stokes 1967 : 192–8).

In this article, Stokes also carried out a similar analysis of variance for British elections from 1950–6. He found that British politics were even more "nationalized" than those of the United States. He wrote (p. 91) "Britain may well provide an extreme case of the nationalization of political attitudes in the Western World; probably it is unique among nations which elect the national legislature from single-member constituencies." Moreover, this is not a recent state of affairs. British elections in the 1890s had distinct overtones of nationalization.

There are then two dimensions to the nationalization thesis. The first is that, over time, distinctive local or regional forms of political expression are replaced by national ones. This, of course, says nothing *directly* about the role of place in producing political expression. But territory as a basis to political expression, however it is produced, is in decline. The second dimension to the nationalization thesis does concern the declining significance of place *in toto*. Over time, elections are increasingly "decided" at levels other than the local one. The local or district effect in elections is more and more a residual one. Taking the two dimensions together, therefore, not only is there a presumptive decline over time in the importance of "primordial" territoriality, there is also a decline in the political significance of place in its wider sense.

Discovering place?

"Revisionism" over the nationalization thesis began in the late 1960s. In the United States this was associated most of all with the work of Burnham (1965), who argued for the importance of "the time dimension and social contexts" in explaining political participation and partisan support. In particular, nationalization was specific to the historical formation of certain electoral coalitions (e.g., the New Deal) rather than a transcendental social process. For Europe, Rose (1970) drew attention to the many political differences that existed between the constituent parts of the United Kingdom. Hechter's (1975) research on the ethnic–territorial cleavages in the British Isles reinforced this point of view. Of course, in France the tradition of electoral sociology had never completely receded, and a geographical perspective on political expression was still prominent (Goguel 1983). Likewise in Italy, political expression was often seen as taking a regional pattern (e.g., Stern 1975; Sani 1976, 1977).

Intellectual revisionism did not take place in a political vacuum. Events in Northern Ireland after 1969, the rise in support for Plaid Cymru in Wales and the Scottish National Party in Scotland after 1966, the growth of separatist movements in Spain, France, and other countries after years of quiescence, Nixon's "Southern strategy" in the US, and the continued and increasingly obvious inability of the nationalization thesis to make sense in such diverse settings as Norway, Finland, France, Italy, and Canada, all encouraged a rethinking of conventional wisdom. Rokkan, a major figure in political sociology and coauthor with Lipset of a "flagship" article in the orthodox tradition, became one of the major critics of the nationalization thesis – though it is important to note that even in the famous article with Lipset, Rokkan commented favorably upon the early school of French electoral geographers who "were deeply conscious of the importance of local entrenchments and their perpetuation through time" (Lipset & Rokkan 1967 : 54). Even Stokes joined the ranks of those skeptical about the "inevitability" of total nationalization. In his book with Butler (Butler & Stokes 1969) he tried to account for "weakening of the class alignment" in British politics by reference to "regional and local variations" from the national "norm."

By today, a considerable amount of evidence has been collected to demonstrate the limits of the nationalization thesis in both its aspects. Evidence is of several types. Some concerns the persistence and/or growth of local and regional forms of political expression. Other evidence suggest that nationalization in the sense of an increasing uniformity of political response is not a marked feature of either North American or Western European countries. Of course, this is "strong" evidence for the continued importance of place in national politics. Political behavior could be based upon place-structured processes even without idiosyncratic political expression or with uniform national patterns of response (see Chs. 3 and 4 for discussion of "outcomes" and "causes").

NATIONAL POLITICAL ALIGNMENTS?

National party systems, in the "modern" sense of sets of political parties "competing" for votes and seats, date from the last decades of the 19th century. It is to such party systems that the nationalization thesis is usually applied. But what evidence is there for the persistence or growth of local and regional patterns of support for particular parties rather than nationalization?

For Britain, Urwin (1982a) and others (Miller 1979, Dunbabin 1980, Crampton 1984, Johnston 1985a) have shown persisting and increasing geographical cleavages in voting behavior. However, varying numbers of candidates and uncontested seats present problems in mapping British vote distributions, as does the necessity to adjust for changes in constituency boundaries. Boundary changes mandate examining groups of elections between revisions rather than all elections. To meet the problem of varying competition, the median vote in each constituency over a group of elections can be computed.

Urwin (1982a: 47–50) presents a series of maps showing the median Conservative vote for the periods 1885–1910, 1918–35, 1950–70, and 1979 (Fig. 6.1). In these maps there is no weighting by population size, and city constituencies have been grouped at an all-city level. But as Urwin points out (p. 50), a "broad, century-old pattern in voting support" is apparent. The south of England has always been a Conservative heartland. Indeed it may have become more so since World War II (Dunbabin 1980: 265, Crampton 1984). Moreover, the Conservative vote has tended to decline with increasing distance from the southeastern core. This is particularly true with respect to Wales and Scotland. In terms of change, there has been a decline in support for the Conservatives in northern English industrial areas and in Scotland as a whole since 1950. Overall, therefore, the Conservatives have been and are the party of the "core." The reverse is true of the other parties, particularly the Liberal, Labour, and the Nationalist parties. They are the parties of the "peripheries."

Urwin (1982b) provides a similar kind of analysis for Germany. He shows clearly how marked were the regional and local patterns of support for regional and national political parties prior to 1933. Since the end of World War II, the major parties – the Social Democrats (SPD) and the Christian Democrats (CDU/CSU) have established their own regional bases of support. CDU/CSU's greatest support comes from the Catholic agricultural areas of the southwest. The spread of the SPD's vote is dependent upon the distribution of industry and population density. It has also failed to mobilize support in the more Catholic districts. But regional parties no longer have the levels of support they acquired in the years before the arrival of Hitler. The electoral system and the "coalition" nature of the CDU and SPD have essentially removed their influence (Urwin 1982b: 222–6).

For France, the most recent edition of the classic essay on the *Géographie des élections françaises sous la IIIe et la IVe République* (Goguel 1970) provides evidence for what is termed "habitual" support for specific political parties. At the risk of considerable oversimplification (Derivry and Dogan 1986), this more complex pattern is represented in terms of two maps, one showing areas of "traditional" right-wing support and the other showing areas of "traditional" left-wing support. In fact, the geography of party support has become rather less marked over the past 20 years as the "center" has grown in French politics at the expense of the extremes (Pierce 1980, Jaffré 1980). But it is startling to see, for example, the pattern of support for the French Communist party (PCF) in the parliamentary election of 12 March 1978 (Fig. 6.2). The PCF is very weak in certain areas, especially in the west, the southeast Massif Central, and Alsace. It performs much better in the north, on the northwestern edge of the Massif Central, and along the Mediterranean coast. French presidential elections have also maintained a locality-based pattern of political expression (Johnston 1982).

The success of the Socialist party in 1981 marks perhaps the greatest extent of nationalization in recent French politics. The Socialist party (PS) and its Leftist Radical allies have taken root all over France. Only rarely have its recent results fallen outside a 20 to 30 percent spread (e.g., for 1978, Fig. 6.3). If very recent performance is any guide, however, and the 1984 European elections did indicate an improvement in the fortunes of the right, "the rise of the centre" may yet see "the revenge of the extremes" and a return to Goguel's classic geographical pattern (Parodi 1978).

In Italy the division between left and right can be viewed, as in France and Britain, as geographical rather than functional and national. An important convention of Italian electoral sociology has been that there are four electoral zones or regions in peninsula Italy in which different patterns of political dominance are visible (Capecchi & Galli 1969; Brusa 1983, 1984b). Zone I, the "Industrial Triangle" covers northwest Italy and includes Piedmont, Liguria, and Lombardy. There is a socialist political tradition in this region but it is also a region in which the Christian Democrats do well. Zone II, the "White Zone" covers northeast Italy and includes the provinces of Bergamo and Brescia in Lombardy, the province of Trento, the province of Udine, and all of Veneto except the province of Rovigno. This is a region with a well-established conservative political tradition. Zone III, the "Red Zone," covers central Italy and includes the provinces of Mantova, Rovigno, and Viterbo; the whole of Emilia except for the province of Piacenza; Tuscany; Umbria; and the Marches except for the province of Ascoli Piceno. This is a region in which the Italian Communist party (PCI) is strongly entrenched. Its support comes from the countryside as much as or more than the cities. Zone IV – the "South" – includes the provinces of Ascoli Piceno; Lazio, with the exception of the province of Viterbo; Campania; Abruzzi e Molise; Apulia; Basilicata; and Calabria. This region is the most varied and has undergone the most change, socially and politically. The Christian Democrats continue to dominate, but in the 1960s faced an improved performance by the PCI in certain areas (Capecchi & Galli 1969). Others parts of Italy – Sicily, Sardinia, and the Val d'Aosta and Friuli on the northern peripheries – have more complex alignments, including regional parties.

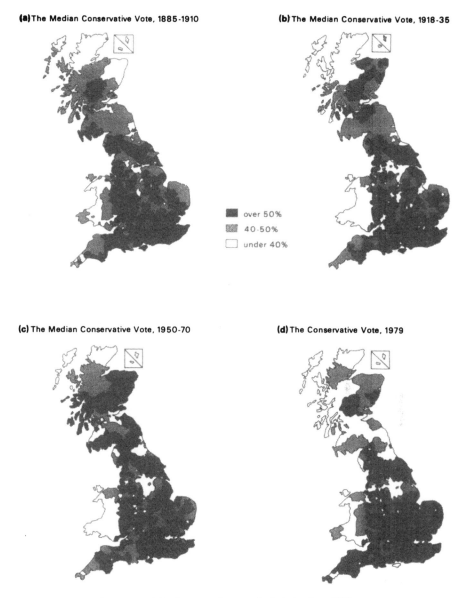

(a) The Median Conservative Vote, 1885-1910

(b) The Median Conservative Vote, 1918-35

over 50%
40-50%
under 40%

(c) The Median Conservative Vote, 1950-70

(d) The Conservative Vote, 1979

Figure 6.1 The Conservative vote in Britain since 1885.
Source: Urwin (1982a : 48–9).

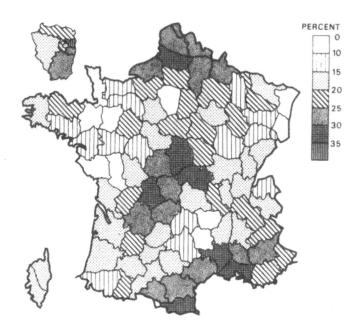

Figure 6.2   The Communist vote in France, 12 March 1978.
Source: Jaffré (1980 : 65).

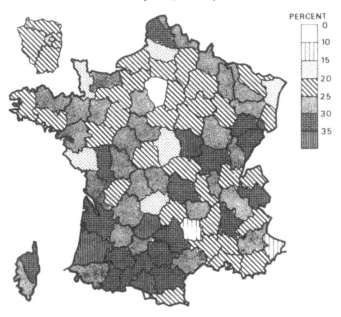

Figure 6.3   The Socialist and Leftist Radical vote, 12 March 1978.
Source: Jaffré (1980 : 65).

Pavsič (1985) argues that this pattern has persisted over the period 1953–83. The only significant shifts have been the increasing support for the Christian Democrats in the south, the persisting regionalization of the Communist vote, and the much increased support for the Socialists in the south. Most of the smaller parties remain regionalized. Pavsič concludes her study of party performance by province by election by emphasizing that "the homogenization of the geographical distribution of votes does not appear to be a general characteristic of the Italian political system" (p. 97, my translation).

In the United States, the "South" is the only region that is usually isolated for its particularity. The rest of the country is regarded as in the "mainstream" of nationalized politics. Even the South is now apparently undergoing nationalization (Beck 1977). Certainly if one examines electoral data at a gross scale, there seems to be a split between Democrats and Republicans in all regions that is distinctly closer than that of earlier electoral history (Polsby & Wildavsky 1980:6–7). This is particularly true of presidential elections, if not of congressional ones (Claggett et al. 1984:86–8). Burnham (1982) has pointed out that each of the major phases of American electoral history has had a distinctive geography (see also Archer & Taylor 1981). From 1932 until the 1960s, the "New Deal System" replaced the stable regionalism of the previous period with an unstable coalition of regional interests. More recently, the Republican party has made major inroads into the previously monolithic Democratic South (Burnham 1982:110–15). At the same time, however, a local pattern of one-partyism has developed as incumbents have insulated themselves from electoral competition (Burnham 1982:207–16). Thus, although nationalization appeared to have triumphed at the regional scale, although this is now under question (Claggett et al. 1984), its reality is certainly called into question at the local scale and since the 1970s.

In Canada, trends have been more straightforward. Territorial cleavages there, at both regional and local scales, are neither "a legacy of the past" nor have they decreased over time. A considerable amount of empirical research has verified the impact of territorial cleavages on electoral behavior. In a study of party voting between 1945 and 1970 in 17 countries, Canada ranked among the least nationalized, while the United States was the most nationalized (Rose & Urwin 1975). However, the 1984 general election produced a nationwide Conservative sweep and the first truly national government since 1957. The stability of this remains to be seen.

In Canada the most distinctive geographical unit is Quebec. Since before 1900, Quebec has provided solid and frequently overwhelming support to the Liberal party in federal elections (Fig. 6.4). Conversely, Quebec has been largely a wasteland for the federal Conservative party except in 1957 and 1984. Elsewhere in Canada there are similar, if less extreme, biases towards one party rather than others (Ley 1984). One commentator has gone so far as to refer to the history of Canadian national politics as a process of "provincialization" (Johnston 1980:173).

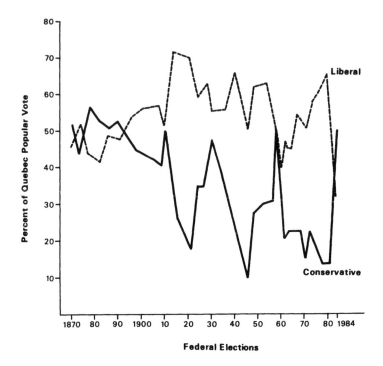

Figure 6.4 Proportion of the Quebec popular vote received by the Conservative and Liberal parties in federal elections, 1870–1984.

Throughout Western Europe and North America, therefore, local and regional patterns of political competition and dominance have persisted, and in some cases even strengthened. It is hard to accept that these are simply exceptions to a rule, or temporary setbacks on the road to nationalization. To the contrary, they seem to indicate a fundamental problem with the nationalization thesis. The image of increasing national social and political homogeneity and social mix upon which it rests is illusory.

UNIFORMITY OF RESPONSE?
The second aspect to nationalization concerns the increasing uniformity of electoral response over time rather than convergence in the levels of partisan support. Stokes (1965, 1967) described nationalization this way. To him, nationalization involved increasing similarity over time of all subunits of the electorate in response to national political forces. This is sometimes called "uniform national swing." It refers to the similarity of response rather than absolute level of partisan support. A set of districts characterized by a high degree of uniformity in swing may be quite different in their levels of partisan support.

Claggett *et al.* (1984) have repeated Stokes's analysis of nationalization as a

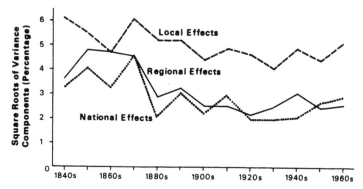

Figure 6.5 Square roots of the variance components for national, regional, and local effects, US congressional elections, 1842–1970.
Source: Claggett *et al.* (1984: 88).

uniform response to political forces using the same US congressional elections (1842–1970) adopted by him. They partition the total variance into three effects: the *national effect*, the election sum of squares, measures the amount of uniform movement across subunits; the *regional effect*, the election–region interaction sum of squares, measures the amount of uniform movement of counties within regions; and the *local effect*, the election–county (within region) interaction sum of squares, measures the idiosyncratic movement of counties within regions and the nation. These three effects for the Democratic percentage of the congressional vote from 1842–1970 are reported in Figure 6.5.

Rather than revealing an increasing national effect, Claggett *et al.*'s (1984) analysis reveals a persistent, if mildly decreasing, local effect as the most important one. Indeed, far from increasing, the national effect declines

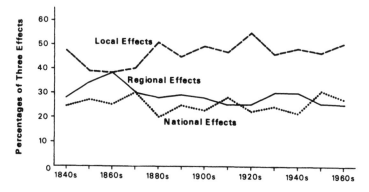

Figure 6.6 Normalized variance components for national, regional, and local effects, US congressional elections, 1842–1970.
Source: Claggett *et al.* (1984: 89).

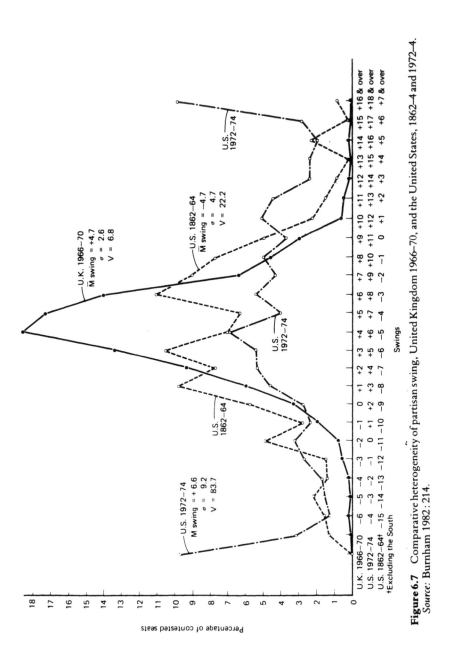

Figure 6.7 Comparative heterogeneity of partisan swing, United Kingdom 1966–70, and the United States, 1862–4 and 1972–4.
Source: Burnham 1982 : 214.

dramatically after the Civil War and remains low thereafter. Stokes, of course, based his conclusion of increasing nationalization in vote choice entirely on the slight *decline* in the local effect.

If the data are adjusted to take into account the variation in total variance over time, the *relative* contribution of the three effects can be assessed more reliably. "Normalizing" the data in this way provides even stronger evidence against a nationalization process (Fig. 6.6). Now the local effect *increases* slightly rather than decreases and the national effect remains low and stable.

An alternative approach involves examining the heterogeneity of partisan swing to see whether or not there is a clustering around a national "norm." Burnham (1982:214–15) has done this for two election periods in the US (1862–4 outside the South and 1972–4), and one in the UK (1966–70). The 1972–4 US swing is vastly more heterogeneous than either the 1966–70 swing in the UK or the 1862–4 American swing (Fig. 6.7). There is again, therefore, little support for the idea that congressional elections are increasingly national events. To the contrary, the diversity of swings supports the idea that congressional elections are *decreasingly* national.

As Burnham's data suggest, swing in Britain has been much more uniform nationally than in the United States. This has led to one of the major paradoxes of British electoral sociology: for swings to be uniform across a set of constituencies with varying initial levels of partisan support, voting behavior must differ substantially from seat to seat (Butler & Stokes 1969:303–12, McLean 1973, Johnston & Hay 1982). Butler and Stokes have proposed a simple model to reconcile the national uniformity of swing with heterogeneous voting behavior (Butler & Stokes 1969:303–12). This model assumes that voters differ in terms of their sources of information, and these differences translate into differences in voting behavior. Three groups of voters are specified, with national, local, and mixed sources of information. National voters obtain their information from the mass media, and show a uniform propensity to shift parties regardless of the dominant political complexion of their constituency. Local voters, in contrast, rely on personal sources of information, such as friends and acquaintances. These sources are strongly influenced by local political complexion. Where a single party is dominant, it is argued, local sources will heavily favor the dominant party, and those who might vote otherwise are persuaded to switch to the stronger party. Supporters of the dominant party will also be reinforced in their predispositions. Thus, local processes generate a "neighborhood effect" that runs counter to the operation of the national effect to produce the uniform swing. The national uniform swing is "rescued" from tautology by proposing a limited contextual effect.

But as Bodman (1982) has argued, the concept of swing has a number of basic limitations in addition to its paradoxical relationship with nationalization. The first limitation is that the concept can only be applied in its present form to swings between the two major parties. As electoral contests have become more complex, its utility has decreased. Another weakness is that swing summarizes the joint effects of a variety of processes. It does not provide the possibility of distinguishing the direct transfer of votes between the two major parties from other processes: the movement to third parties, the circulation of voters between major parties, nonvoting, changes in the demographic characteristics of the electorate as a result of death and enfranchisement, or the alteration of the electorate through migration (Bodman 1982:36). Finally, swing measures only the *change* in the relationship between the two major parties. Quite different sets of party performance can produce the same swing. For example, a high swing to Labour may be the result of a poor Conservative performance and an average Labour one. Alternatively, a high swing to Labour may result from an average Conservative performance and an above-average Labour one (Bodman 1982:39). Exactly what a uniform national swing indicates, therefore, is open to question. What is clear is that the connection between uniform national swing and nationalization is tenuous at best.

The question of uniformity of response in national elections as an indicator of nationalization has also been investigated for other countries. Not surprisingly, because the United States and the United Kingdom have usually been referred to as the "most" nationalized political systems, results of research elsewhere suggest even more limited degrees of nationalization. Jackman (1979), for example, partitioned swings in the popular vote in Canada into national, provincial, and riding (district) effects, following the lead of Stokes. He found that Canadian electoral behavior was less nationalized than seemed to be the case in Britain and the United States. He also found that provincial effects in Canada were much stronger than state effects in the United States, and that they were increasingly so over time. He concludes (p. 141) that even in Anglophone Canada "there is little evidence for the notion that Canadian parties have operated as nationalizing agents in the twentieth century or that Canadian voters behave as members of a nationally integrated community." (See also Ley 1984.)

Browne and Vertz (1983) provide evidence from West Germany against the increase in uniformity of response in national elections over time. Using Stokes's model they began by expecting considerable nationalization of shifts in support for West German political parties over the period 1949–76. What they found, however, is that for all parties together and for the three major parties – the SPD, the CDU/CSU, and the Free Democrats (FDP) – separately, "there is no generalized nationalization of electoral forces acting upon constituency voting" (Browne & Vertz 1983:91). This is particularly marked for the individual parties. Indeed, they refer to them in terms of the increasing "localization" of electoral support. This matches the findings of others who have observed an increase in the importance of local issues in West German elections, the long-term decline of social attributes as powerful predictors of electoral behavior, the increasing importance of local party organizations, the growth of "new" issues (such as those of the Green parties), and the movement of parties away from their traditional social bases (Klingemann & Pappi 1970, Conradt 1973, Baker *et al.* 1981).

A range of recent studies suggests, therefore, that the nationalization thesis in both its aspects – convergence in levels of partisan support and increasing uniformity in response to electoral forces – has little or no support. But if we abandon it, what can substitute for the national-level accounts of political behavior that have accompanied it? There are a number of options, most involve adapting the conventional approaches examined in Chapter 2. One solution, that advocated in Chapter 3, is more radical.

Fixing accounts

A number of approaches have been proposed to deal with the reality of geographical variation that has become increasingly apparent to political sociologists. Some of these are relatively modest. One involves "explaining away" local and regional variation as a function of population composition – for example, certain regions have more of the working class and more Catholics, and so on. Another strategy is that proposed by Butler and Stokes (and others) for fixing the nationalization thesis by positing a separate and inferior local effect – the neighborhood effect.

Some other approaches are more ambitious. Four stand out. One is that proposed by Rokkan and Urwin in recent work, to the effect that a latent "territorialism" has been triggered by what they term the "politicization of peripheral predicaments" (Rokkan & Urwin 1983). Another is that of various "uneven development" theorists. According to this viewpoint – and there are substantial differences within this group – capitalism, industrialism, or modernization has not diffused evenly, but rather has always involved the domination of some regions by others. This process generates a political response from the suffering regions in the form of regional "nationalism" or support for parties opposed to the geographical status quo (e.g., Hechter 1975, Nairn 1977).

The third more ambitious approach focuses on interethnic competition rather than geographical variation per se. The argument runs as follows: modernization produces competition for occupations, rewards, and social roles between ethnic groups as a previous cultural division of labor breaks down. Competitive tensions between groups are manifested in an increasing solidarity within groups and political conflict between groups (e.g., Barth 1969, Hannan 1979, Nielsen 1980). Since ethnic groups usually occupy distinctive "niches," their conflicts take on a geographical form.

The final approach denies that there is a geographical problem at issue. Rather, the problem is "partisan dealignment" and possible realignment around new social cleavages and "new-issue agendas" (e.g., Särlvik & Crewe 1983; Flanagan & Dalton 1984; Dunleavy 1979, 1980). In particular, the class-based cleavage around which most political parties have been organized is "obsolete." Proponents of this viewpoint differ in terms of whether dealignment is a temporary phenomenon, one in which specific political parties are losing support and being replaced by other ones (the realignment thesis), or a permanent shift away from political parties towards "participating" and interest-group politics (the dealignment thesis) (Flanagan & Dalton 1984). On this account, whether emphasizing dealignment or realignment, the breakdown of the nationalization process, however, is only a temporary affair. Either new national parties will emerge (better yet, the old "stable" system will make a comeback), or new national issue-group cleavages will displace the old social-group ones. I shall now examine these arguments in more detail.

POPULATION COMPOSITION
Simply put, this approach proposes that geographical variation in support for political parties is a function of population mix or "composition effects." In particular, social classes and religious groups are not evenly distributed, and much of the regional and local variation in party support is due simply to this uneven distribution of social groups (e.g., Blondel 1963, Alford 1963, Galli & Prandi 1970, LeBras & Todd 1981, Miller & Raab 1977). If there was a more even social mix across electoral districts the nationalization thesis would work.

The problem with this, as Butler and Stokes (1969: 137–44) have pointed out is that, at least in Britain (and the data presented by Galli & Prandi (1970) and LeBras & Todd (1981) suggest also in Italy and France), overall differences in party support between regions are due more to variations in party support by class than to the different class compositions of the regions. For example, at the very general level of "north" (Scotland, Wales, north of England) and "south" (south and Midlands) for the period 1963–6, Labour's support among working-class electors in the north was as high as the Conservative's support among middle-class electors; the amount of "cross-support" in the two classes was much the same. In the south, however, Labour's share of the working-class vote was much less than the Conservatives' share of the middle-class vote; the balance of "cross-support" was very much in the Conservatives' favor (pp. 142–3). The explanation offered (pp. 151–70, 144–50) is that the major processes at work are trade union influence and the influence of local political environments through information flow and "persuasion." This immediately carries the argument away from population composition to place-specific effects.

THE NEIGHBORHOOD EFFECT
As Butler and Stokes (1969: 144) note, "the tendency of local areas to become homogeneous in their political opinions has attracted the attention of many observers and been described in many ways." They propose that the

Table 6.1 Partisan self-image by class in mining seats (Labour) and resorts (Conservatives).

	Mining seats			Resorts	
	Class self-image middle class	working class		Class self-image middle class	working class
Conservative	64%	*b* 9%	Conservative	93%	*b* 52%
Labour	*c* 36%	91%	Labour	*c* 7%	48%
	100%	100%		100%	100%
	b − *c* = −27%			*b* − *c* = +45%	

Source: Butler & Stokes 1969 : 145.

"persuasive effects" of dominant local opinion in counteracting national class influences are the major processes at work in creating homogeneous local areas (see also Ennis 1962, Cox 1968). As mentioned earlier, Butler & Stokes (1969:303–12) identify a particular group, those reliant on local sources of information, as the ones responsible for local homogeneity. They are the "switchers" from national class position to "local" outlook.

At first sight the evidence for neighborhood effects is strong, both from survey evidence and aggregate data (see Taylor & Johnston 1979:221–64). Butler & Stokes (1969:146–50) themselves provide persuasive data. Using survey data they show, first, the relationship between the proportion that the middle-class constituted in each constituency and the proportion Conservative among middle-class voters who supported one of the two major parties in each constituency (Fig. 6.8). Secondly, they show a similar relationship with proportion working-class and proportion Labour (Fig. 6.9). The more dominant a class is in a constituency, the greater the proportion of its members that vote along class lines. They also demonstrate that in constituencies that are the bastions of a major party, the major party receives a high proportion of votes from the class whose party is the minor one (Table 6.1).

Further analysis of Butler and Stokes's survey data, however, suggests that the neighborhood effects as usually conceptualized are in fact a very minor determinant of voting behavior (Tate 1974). Tate used both individual and contextual variables to discriminate between Labour and non-Labour voters. The individual variables accounted for over 32 percent of total variance with occupation the best discriminator. The contextual variables accounted for only 6.4 percent, with class composition of the electorate the best discriminator (p. 101). One problem with Tate's approach, however, is that the procedure used assumes that context involves an *additive* increment or decrement. In Hauser's (1970:662) words, the assumption is "that the individual predictor variables have exactly the same

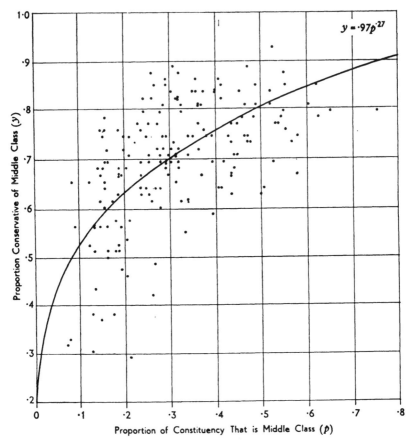

Figure 6.8 Conservative proportion of middle-class support for the two major parties in Britain by proportion of constituency that is middle class. Based on data from 184 constituencies in which the National Opinion Polls interviewed more than 120,000 respondents from 1963 to 1966. The analysis shown here is limited to electors who supported the Conservatives or Labour. "Middle class" is defined as occupational grades A, B, and C1.

Source: Butler & Stokes (1969: 147).

influence within every group [context] and the unique effect of the group is just an additive increment or decrement." There is no role for empirical analysis of context–individual interactions in this kind of model (see McAllister 1987, Agnew 1987b).

More fundamentally, perhaps, the concept of a neighborhood effect as presently defined rests on a number of dubious premises. Bodman (1983) points out that the Butler and Stokes data reported in Figures 6.8 and 6.9 cannot be manipulated to reveal the conversion of minority local voters to the majority view. The pattern in the data and the purported process cannot be reconciled. This directs attention to the model's assumptions. One assumption is that voters are equally susceptible to conversion, irrespective

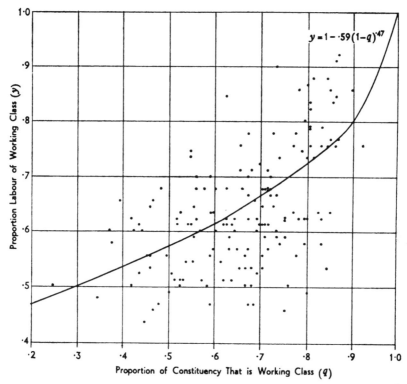

Figure 6.9 Labour proportion of working-class support for the two major parties in Britain by proportion of constituency that is working class. Based on data from 184 constituencies in which the National Opinion Polls interviewed more than 120,000 respondents from 1963 to 1966. The analysis shown here is limited to electors who supported the Conservatives or Labour. "Working class" is defined as occupational grades C2, D, and E.
  Source: Butler & Stokes (1969:148).

of strength of party identification. Other equally unlikely assumptions include equal probability of sending and receiving partisan messages, random mixing of minority and majority voters within constituencies, and the premise that there are clearly distinguishable local and national voters (Bodman 1983:130, Pelling 1967, Johnston 1981). To the extent that there are processes of political persuasion at work these are much more likely to be at a *very* local scale rather than that of the constituency. The patterns in Figures 6.8 and 6.9 may be much better accounted for by variations in party organization; different rates of abstentionism; or place-related differences in political outlook (Bodman 1983; Johnston 1981, 1985b; Johnston *et al.* 1985).

  But perhaps the neighborhood effect can be rescued – after a fashion. Butler and Stokes (1969:151–70) propose that union membership is a form of interpersonal influence which strengthens class cohesion among manual workers. Consequently, in areas where union membership is more extensive this reinforces the propensity to vote Labour. The problem with this is that unionization is not a typical or exclusive attribute of manual workers. As Dunleavy (1980:370–1) demonstrates, there is no simple relationship between unionization and class defined in occupational terms: "There are very large groups of manual workers who are barely unionized, and there are similarly a substantial minority of non-manual employees working in industries which are amongst the most unionized of any." Of course, many union members, whatever their occupations, vote Conservative or for a third party.

Butler and Stokes (1969:145) warn their readers when they embark on their discussion of the "local political environment" that the differences between local areas displayed by their survey data "ought not to be assigned too readily to any single explanatory principle." Yet this is what they do. In their attempt to rescue uniform national swing (and the nationalization thesis), they prematurely committed themselves to the neighborhood effect as *the* principle for explaining away regional and local variation. This way, national class cleavages can be kept as the major explanation for voting behavior and "a number of variations on the theme of class" can be thrown in to account for local and regional differences.

THE "TERRITORIAL DIMENSION"

In a recent series of articles and books, the late Stein Rokkan and his associate Derek Urwin have proposed a "territorial" approach to politics which rests on a center–periphery model of state building (Fig. 6.10). They stress the "multidimensionality" of center–periphery relationships (Rokkan & Urwin 1983:16). There are three sets of relationships: military–administrative, economic, and cultural. Peripheries can be conquered by centers in all or perhaps only one way. But changes in one type of relationship have effects on others. For example, increasing economic dependence affects cultural distinctiveness.

Rokkan and Urwin focus on the territorial structuring of Europe. Their point of departure is a "conceptual map" of Europe which identifies a central European city belt, the core areas of state building, and peripheries which have resisted cultural incorporation into modern states. The objective is to construct a typology of "peripheral predicaments": political reactions of peripheries to processes of subordination and incorporation. A distinction is made between "subject peripheries," such as Wales, and "interface peripheries," caught in the cross-fire between different centers, such as Alsace.

Urwin (1980, 1982a) has adapted this model to an examination of modern British electoral politics. Urwin's major point is that until the 1880s, British politics was local and regional. With what Urwin (1980:232) terms "social modernization" (industrialization, the decline of agriculture, urbanization, secularization etc.), and the "consolidation of formal mass democracy" (franchise extension and the supervision of elections), this changed. Class as

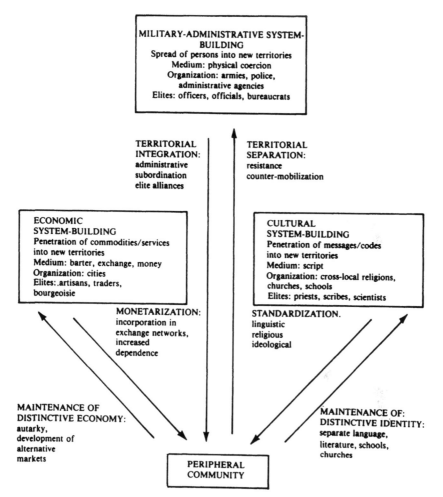

Figure 6.10 Rokkan and Urwin's model of interaction and resistance within large-scale territorial systems.
Source: Rokkan & Urwin (1983:15).

a national phenomenon rose in importance. Yet class had to conform, so to speak, to the established relationships between center and peripheries. Rather than displacing territorial politics, therefore, new class-based alignments fed into existing regional ones (Urwin 1982a:47). The expansion in the 1970s of regionally based parties such as Plaid Cymru in Wales and the Scottish National party (SNP) in Scotland represents the resurgence of territorial and, perhaps, other interests no longer met by the dominant, statewide parties.

The great appeal of this approach lies in its reliance on a historical analysis of state building, and the link between local and regional history and contemporary patterns of political expression. The message is clear: there is nothing *inevitable* about nationalization.

There is a problem with the approach, however. The major geographical units that are basic to the analysis are "centers" and "peripheries." But this is to substitute in any given state two or three units for the present singular unit: the entire state territory. This does not take us very far in explaining geographical variation *within* centers and peripheries. In the absence of any other suggestions, we are thrown back solely on the neighborhood effect.

UNEVEN REGIONAL DEVELOPMENT

This approach, or rather set of approaches, is a variation on the theme of territorial politics, but one which emphasizes the overwhelming importance of economic *underdevelopment* in the creation of peripheries (e.g., Hechter 1975, Nairn 1977). In Nairn's version, uneven development is a product of imperialism in which center–periphery relationships in states such as Britain are one of the last links in a global chain of metropolis–satellite relationships. Taking this at face value, uneven development need have no ethnic connotations whatsoever (Sloan 1979). To Nairn (1977: 353), however, as capitalism spread, the "ancient social formations" it assaulted fell apart along the "fault lines" that defined them. These were lines of nationality and religion rather than class (Nairn 1977: 353).

Hechter's (1975) version is a model of "reactive ethnicity." Economic development typically favors core regions and core ethnic groups. Resurgences of ethnic solidarity and protest are viewed as reactions of culturally distinctive peripheries against exploitation by the center. This exploitation takes the form of a "cultural division of labor" in which the population of a periphery engages in occupations and social roles that are the most rewarding to the center and least profitable to the periphery (Hechter 1975).

The cultural division of labor, as noted by Ragin (1976), is akin to the overlapping cleavages element in Lipset and Rokkan's (1967) model of political mobilization: economic and cultural cleavages are mutually reinforcing. To Lipset and Rokkan, of course, this was never a permanent relationship. Cultural cleavages eventually wither. But Hechter's reasoning follows the same path. The cleavages are *functional* to the perpetuation of the status quo.

The major problem here, as it is with Nairn's argument, is the *relative* importance of economic and cultural cleavages in generating nationalism or political distinctiveness. Orridge (1981: 188) puts this in a nutshell: "Does uneven development alone create the sense of separateness or does it need a strong pre-existing sense of distinctiveness to work on, thus accounting for the absence of a sense of separate nationality in Southern Italy or Northern England?" This is a critically important question. If prior ethnic identity is necessary, and uneven development is insufficient by itself to produce political distinctiveness, then it becomes difficult to apply this approach except to cases where ethnic distinctiveness is always present.

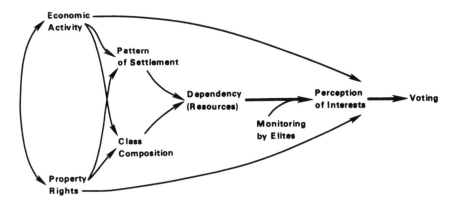

Figure 6.11 Brustein's mode-of-production model of French political behavior.
Source: Brustein (1981 : 369).

However, there is clearly a sense in which the pattern of economic growth within states, uneven development, is related to the pattern of political expression. It is the singular contribution of Brustein (1981) to have made this clear. He argues that different regional "modes of production" have produced different "constellations of interests" in western as opposed to Mediterranean France. This economic difference is held to account for the different modes of political expression in the two regions: right-wing voting in western France and left-wing voting in Mediterranean France. (Though he does not cite him, Brustein is following a path trod previously by Fox (1971).) Unfortunately, Brustein makes no attempt to link his regions into a wider French or European economy. They are worlds unto themselves. Yet it is not clear how independent they are, or why these specific regions were chosen except because of their *political* rather than their economic characteristics.

More fundamentally, Brustein juxtaposes a holistic conception of economy with an individualistic model of voting behavior (Fig. 6.11). The mode of production concept provides a paradoxical backdrop to the image of "voters as strategic actors who vote for those candidates or programs that will provide them with the greatest benefits" (p. 367). Can a mode of production allow *strategic* action? Another limitation is the relentless economic determinism of the argument. Questions of ideological commitments and religious divisions as being anything other than superfluous are summarily dismissed (Hunt 1984). Where is the scope for *contingency*?

The concept of uneven development is a fruitful one for examining geographical variation in political expression. As yet, however, it has either been mixed with limiting concepts of ethnicity or pursued with a relentless economic determinism and eclectic ontology.

ETHNIC COMPETITION

This approach concerns itself with the rise of political differences based upon increased ethnic solidarity. In this respect it is not unlike Hechter's reactive ethnicity model. However, unlike that model, the ethnic competition model associates solidarity with situations in which there is an equalizing of development rather than uneven development. It is as the cultural division of labor *breaks down* that ethnic competition for jobs and social roles ensues. As long as different groups occupy distinct sectors of activity or "niches," interethnic relations are stable. Instability occurs when the groups compete for the same rewards (Barth 1969, Nielsen 1980).

This approach appears to work well in explaining the rise of the Flemish movement in Belgium since World War II (Nielsen 1980). However, it is unclear how generalizable it can be. Only a considerable stretch of the imagination could interpret Irish or Basque nationalism this way. But perhaps it can be applied to settings where immigrant minority groups are struggling for recognition or where a resident group is faced with economic competition from an immigrant group on terms to the advantage of the immigrants (e.g., the oil boom in northeast Scotland).

Again, from the general perspective of political sociology this is another "fix it" strategy to cope with exceptions to the *rule* of nationalization. In Belgium at least, it copes with the ethnic revival among Flemings in the more economically developed cantons. Nationalization has been the lot of their peers in the "backward areas."

PARTISAN DEALIGNMENT

Over the past several years, students of political party systems have noted the emergence of new parties, increases in partisan volatility at the aggregate and individual levels, and declines in levels of party identification. It is these changes rather than disenchantment with the nationalization thesis that has led to criticism of conventional theorizing. Indeed dealignment is itself viewed as a *national* phenomenon. Typically, the declining association between occupational class and political alignment is seen as the key feature of partisan dealignment (Rose 1974, Francis & Payne 1977, Crewe *et al*. 1977, Franklin & Mughan 1978, Särlvik & Crewe 1983, Pederson 1983, Dalton *et al*. 1984). This trend is usually put down to the growth of "postindustrial society" and the decline of class conflict as occupational inequalities have lessened and the welfare state has provided economic and social security.

There are several problems with this approach. The first is that dealignment presupposes that there *was* a set of national partisan alignments based on national class division that could be dealigned. An earlier section of this chapter demonstrated the frailty of this assumption. A second problem lies in the choice of explanation for recent dealignments. One plausible explanation is that one or several new cleavages have arisen or grown in political significance such that they cross-cut a previously dominant cleavage (such as occupational class), reducing its impact and creating cross-pressures on certain segments of the population (Rae & Taylor 1970). Dunleavy (1979, 1980) argues persuasively for increases in state employment and a public–private cleavage in modes of consumption as examples of such cross-cutting cleavages (see the next section).

At the heart of the dealignment thesis lies a reinterpretation of challenges to conventional wisdom in congenial terms. Rather than facing up to the bankruptcy of the nationalization thesis and its single-minded obsession with national occupational class cleavages, dealignment theorists reinterpret *dramatic* shifts in employment patterns and cleavages in consumption patterns in terms of their conventional national categories.

Dealignment theorists do differ, however, in terms of the consequences of dealignment they identify for individual parties and party systems. According to one view the present instability is a passing phase, a feature of transition from one stable alignment to another. A contrasting viewpoint argues that parties have become obsolete. This position is particularly popular with commentators on American politics. Some wags might argue that obsolescence has been a feature of American parties for most of this century! Both points of view, however, rest on an image of historical stability. Yet to the extent that there ever were stable national alignments in most countries based on occupational class, they lasted only from 1950 until 1965. Periodic "partisan realignment" and dealignment have been features of political systems for many years (Key 1955, Burnham 1982, Urwin 1980). "Instability," then, may be the *normal* condition of parties and party systems. It is not difficult to conclude that much writing on dealignment represents a crisis for conventional political sociology rather than one of political systems.

Place perhaps?

None of the approaches proposed to account for the failures of the nationalization thesis is either conceptually or empirically satisfactory. They all fail to meet the criteria of conceptual adequacy identified in Chapters 2 and 3 and the empirical requirements demonstrated earlier in this chapter. The discovery of place has been stillborn, as it were. Rather than following through on doubts about the nationalization thesis and conventional modes of theorizing, commentators have remained committed to frames of reference that preclude a satisfactory response. However, within some of the categories of response there have been *signs* of a shift *towards* the framework put forward in Chapter 3.

Urwin's (1980, 1982a, 1983b) work on the historical political sociology of Britain and Germany, for example, highlights the issue of geographical variation. Rather than a "secondary" issue, the geographical distribution of support for specific political parties and movements assumes a central position. This is seen most clearly perhaps in Urwin and Aarebrot (1981). A central question there is: "What happens in a territorially segmented society where for each region there is not just a different class structure, but a different class relationship with voting?" (Urwin & Aarebrot 1981 : 264). The answer for Weimar Germany is an incredibly complex pattern of political expression. This supports the idea that the social and the geographical are not independent of each other. Indeed, "membership space [the social characteristics shared by individuals] and territorial place [location and occupation of territory] go hand in hand" (Urwin & Aarebrot 1981 : 272).

Brustein's (1981) *objective* in linking regional economic characteristics and political expression in France is also resonant with those of the place perspective. His work is indeed very much in the tradition of French electoral sociology which is one of the sources for this perspective. Perhaps the major drawbacks of Brustein's research lies in its obsession with *generalization* after the fashion of Siegfried rather than *causal explanation* after the fashion of Chapter 3. This leads him into the positivist trap of assuming that there is dependence of individual behavior on macrosociological external causes that when correlated explain political behavior. This view is quite at odds with the one proposed in Chapter 3 (see also Agnew 1981b).

Perhaps the closest approach to that of the place perspective is found in Dunleavy's (1979, 1980) writing on the impact of state intervention and the growth of consumption cleavages on political alignment (see also Johnston 1983). The model of political alignments which Dunleavy adopts stresses the *social* character of interest perception and value formation. They are the product of structuration. Thus, local processes of service provision, and conflict over provision, particularly in housing and transportation, have produced in modern Britain significant geographical variation in public as opposed to private provision of services (see also Sharpe & Newton 1984). Different local governments have produced different types of provision of housing, transport, and other goods in response to demands of local electorates. The strength of local Labour party organizations, for example, has come to depend to a great degree on the successes, or failures, of "municipal socialism." At the same time, the Conservative party at the local level has come to base much of its support upon protecting owner-occupied and car-owning constituents from the incursions of public housing and public transport (Dunleavy 1979 : 432–6). *Local* collective interests and demands can be important determinants of contemporary national political alignments.

Conclusion

Chapter 5 made clear the depth of prejudice against the concept of place in modern social science. This chapter has reported data which might lead one to look more favorably upon it. What is clear, however, is that, with a few exceptions, reactions to the palpable failure of the nationalization thesis to hold up under close scrutiny have not led to a revaluation of place. Rather, attempts have been made to plug the holes in conventional modes of theorizing. They are still leaking.

Why is there resistance to rethinking in terms of place? Chapter 5 covered most of the ground. But one point needs stressing: the nationalization thesis is not just an academic position. It has become a political commitment for many academic social scientists. By criticizing the nationalization thesis one is also criticizing a cherished political commitment. That it is probably devoid of empirical validity is, it seems, very much besides the point.

Attention now turns to two "case studies" in which political behavior is examined from the place perspective. The first of these involves Scotland since 1885. The second views American politics since 1880 from the place perspective.

7 *Place and Scottish politics*
Aggregate political behavior

If Ireland was the region of political difference *par excellence* in 19th-century British politics, Scotland has acquired a similar claim to fame in this century. Especially over the past 20 years, Scottish politics has evidently "denationalized" from the standard "British" pattern of the period 1945–64 (Miller 1981). Numerous attempts have now been made to account for this shift. Most follow one or other of the "fixing accounts" described in Chapter 6. The two most favored are uneven development accounts that isolate Scotland's experience as a cultural–economic periphery (e.g., Hechter 1975, Nairn 1977) and population composition plus neighborhood effect accounts that point to the larger working-class element in Scotland's population relative to that of England (e.g., Miller 1981).

The major purpose of the next three chapters is to present evidence from Scotland that is both illustrative and supportive of the place-based perspective on political sociology outlined in Chapter 3. It therefore extends beyond an analysis and account of contemporary dealignment into an investigation of the various ·facets of the place-based perspective. At the same time, however, a major claim is that dealignment cannot be satisfactorily explained unless it is seen from this point of view.

The Scottish case is organized in three separate chapters. These cover three major categories of "evidence" for the place-based perspective outlined in Chapter 4. They provide a useful framework for interweaving place with politics. The first chapter, on aggregate political behavior, provides an overview and attempts an explanation of place-specific patterns of electoral behavior from 1885 to the present. The second chapter focuses on the historical constitution of political behavior in places; it does so by examining the links between the modern social and economic history of, and political expression in, specific places. In a final chapter the links between place and political mobilization in Scotland are identified, especially as these relate to the growth of "new" political movements such as the Scottish National party (SNP). A short conclusion to that chapter summarizes the argument that political expression in Scotland is *intrinsically* geographical. It is not merely the by-product of an uneven development or population composition that draws Scotland away from a national British "norm." Rather, it is the product of political behavior structured by the historically constituted social contexts in which people live their lives: in a word, places.

Aggregate political behavior

ELECTORAL BEHAVIOR 1885–1983

The attempt to relate political behavior, such as voting in elections or referenda, to geographical patterns of social life often begins, and sometimes ends, with the mapping of aggregate outcomes. This is followed by a discussion of the "political maps" in terms of the geographical distribution of occupational groups or religious affiliations. Kinnear (1968) has followed this strategy in providing the only major "ecological" study of British elections over an extended time-period (1885–1966), although Pelling (1967) follows a similar if less explicit approach. Kinnear divides his coverage into three periods that fit Scotland as well as Britain as a whole: Conservative and Liberal dominance from 1885 to 1918, the transitional and unstable twenties and thirties, and finally the Conservative and Labour era from 1935. Four of Kinnear's political maps are reproduced with the focus on Scotland, in Figure 7.1. They are for elections in the years 1885, 1918, 1935, and 1955. A map for 1983 has been added to represent a more recent period of "instability." Each period in modern Scottish electoral history, therefore, is represented by at least one map. This oversimplifies the picture considerably, but it is in the interest of clarifying discussion.

The maps illustrate some important trends in Scotland as a whole and for specific localities and regions. The maps of course reflect the criterion of seats won, not votes cast. This tends to exaggerate the unevenness of party support between different areas; but in the final analysis, "first past the post" politics is determined by seats, not votes.

The general election of 1885 was the first in Britain in which a majority of the adult male population was eligible to vote. The electorate rose from 3 million to 5¾ million, 59 percent of the adult male population of Britain (Blewett 1965:31). New electoral boundaries gave Scotland and some other areas extra seats. This helped the Liberals who established themselves in 1885 as the major party in Scotland. They won 58 out of the 70 Scottish seats, many with large majorities. In Northern Scotland, four "Crofters" defeated Liberal candidates, but sat with the Liberals in Parliament. Thus all but eight Scottish seats were in Liberal hands. This pattern of dominance, though it waxed and waned, remained in place through 1918.

The only part of Scotland in which the Liberals were not dominant in 1885 and which remained Conservative or Liberal Unionist until after 1918 was the southwest, particularly the extreme southwest, Kirkcudbright and Wigtown, Bute, and the more rural Clydeside constituencies, Dunbartonshire and West Renfrew. The conventional explanation of this holds that the Liberals were tainted by their support of Home Rule for Ireland (Pelling 1967). In areas with substantial Irish Catholic immigrant populations, the Liberals were easily labelled as the "Irish" party. By implication, Protestants would support their opponents.

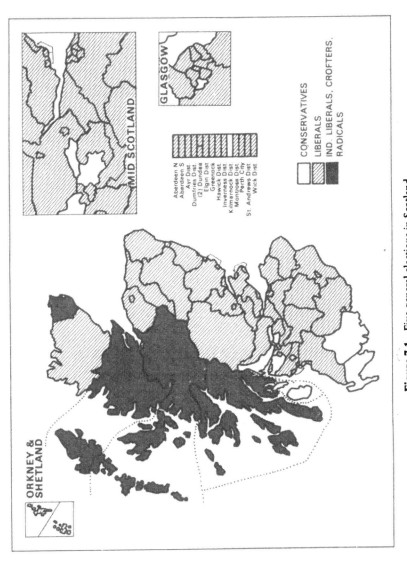

Figure 7.1 Five general elections in Scotland.
Source: Kinnear (1968: 24, 38, 52, 62); *The Times* (London), June 10, 1983: 4–8.
Figure 7.1a (above) 1885.

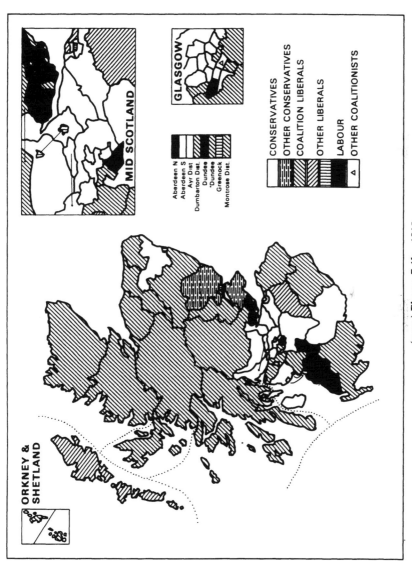

ORKNEY & SHETLAND

MID SCOTLAND

GLASGOW

Aberdeen N
Aberdeen S
Ayr Dist.
Dumbarton Dist.
Dundee
'Dundee
Greenock
Montrose Dist.

CONSERVATIVES
OTHER CONSERVATIVES
COALITION LIBERALS
OTHER LIBERALS
LABOUR
OTHER COALITIONISTS

(cont.) **Figure 7.1b**   1918.

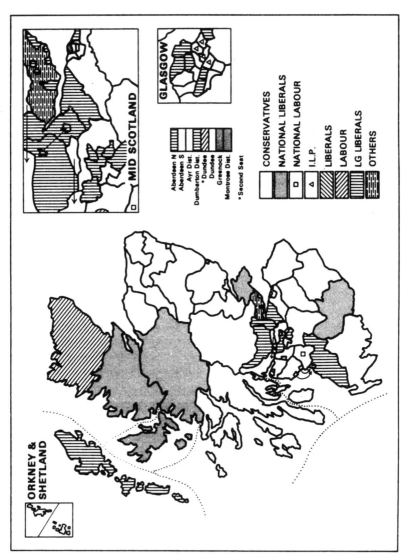

ORKNEY & SHETLAND

MID SCOTLAND

GLASGOW

Aberdeen N
Aberdeen S
Ayr Dist.
Dumbarton Dist.
*Dundee
Dundee
Greenock
Montrose Dist.

* Second Seat

CONSERVATIVES
NATIONAL LIBERALS
NATIONAL LABOUR
I.L.P.
LIBERALS
LABOUR
LG LIBERALS
OTHERS

(cont.) **Figure 7.1c** 1935.

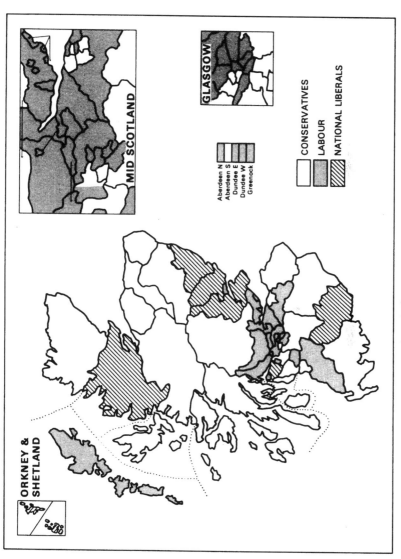

(cont.) **Figure 7.1d** 1955.

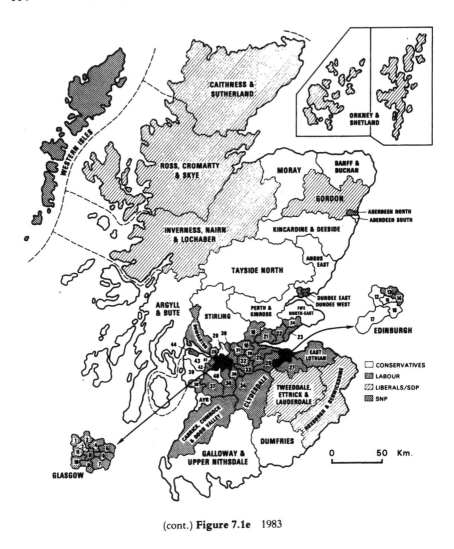

(cont.) **Figure 7.1e**   1983

This argument has considerable merit. The split in the Liberal party in 1886 over Home Rule was largely manifested in a shift to Liberal Unionists and Conservatives in southwest Scotland. But other issues were also at work. In particular, the Liberal party in Scotland in 1885 was a coalition of often opposed groups rather than a monolithic apparatus. Part of its success lay in appealing to different groups with different interests in different places (Savage 1961). But Home Rule for Ireland, and in 1885 at least, the issue of disestablishment of the Church of Scotland, were divisive enough to force a parting of the ways between elements in the Liberal coalition. After 1892,

Liberal–Labour fights provided the Conservatives with their major opportunity in many constituencies (Dunbabin 1980:251).

The Home Rule issue certainly lived on in southwest Scotland for many more years. But after 1918, and especially after 1922, the Labour party began to mobilize electorates across the religious divide. The extension of the franchise in 1918 tripled the size of the electorate. In Scotland, if not elsewhere, this was to the benefit of the Labour party. In 1922 the Labour party won ten of twelve seats contested in Glasgow, and achieved success throughout central Scotland. The Liberal party, as factionalized as ever, disintegrated from 1918 onwards throughout this region. Elsewhere in Scotland the Liberals were increasingly challenged by the Conservatives. Though they maintained a hold on Ross and Cromarty for many years and other rural seats in the north and northeast intermittently, the Liberal hegemony in Scotland disappeared in the 1920s.

The election of 1935 marks the return of predictability after the anomalies and instability of the period from 1918. After 1935 the Conservative and Labour parties held clearly defined groups of seats with homogeneous social composition and divided the more "mixed" seats. The Liberals tended to win only in rural and agricultural seats where they benefited from Conservative–Labour splits. Increasingly, and with the exception of 1955 when they made something of a "comeback," the Liberals lost these seats to Conservatives or, in the case of the "radical" crofters in the Western Isles, to Labour. Only in northern Scotland did they maintain a strong presence.

After 1966 this "classic" 1935 pattern broke down. Beginning in 1970, Scottish politics became a four-way contest. The Scottish National party (SNP), campaigning on a platform of independence for Scotland, has joined the traditional three in the competition for votes. The SNP's efforts have been best rewarded in seats the Liberals had previously lost to the Conservatives, particularly in northeast Scotland. But by 1983 the only two seats the party held, Western Isles and Dundee East, had been safe Labour seats for many years. The SNP, however, has been unable to break into the Labour hegemony in central and southwest Scotland.

Since 1966 the Liberals, singly and in 1983 in alliance with the new Social Democratic party (SDP), have managed to maintain their presence in northern Scotland and expand in the Borders region. In some seats, such as Northeast Fife and Gordon, this may reflect the increased propensity for "tactical voting" when voters are faced with four alternatives, several of which are similar to them. Thus, in a seat such as Gordon in the Northeast, the Liberals have benefited from the votes of those who might have preferred an SNP or Labour candidate to the Liberal, but preferred the Liberal to the Conservative and believed that the Liberal was most likely to acquire the needed votes. But there are limits to this explanation (Kellas & Fotheringham 1976:160). In both the 1974 elections and in 1983, the occurrence of tactical voting seems to have been very limited outside of a couple of constituencies (cf. Butler & Kavanagh 1975 : 276). Conservatives have often been the victims of swings to the SNP. Opinion polls suggest that SNP support is derived from all sorts of party identifiers and socioeconomic categories (Brand & McCrone 1975 : 416). These also suggest that though relatively more working-class identifiers vote SNP than do middle-class identifiers, the major strength of the SNP in the mid-seventies lay with younger voters. In a 1975 survey, almost one-half of those aged 35 or under voted for the SNP. SNP support, even if hardly persistent, let alone loyal, in many places, cannot be explained away by references to tactical or anti-Conservative voting.

From one point of view, the map of Scottish electoral behavior has been tremendously transformed over the past 100 years. Labour has displaced the Liberals in central and southwest Scotland. The Conservative vote in the southwest is much declined. The Conservatives, after a period of resurgence from 1935 until 1966, have gone into decline throughout rural Scotland. They are now threatened by the SNP as well as the Liberals.

There are, however, some continuities. Scotland is on the whole, and by British standards, a "radical" country. The Conservatives have always done poorly in Scotland relative to most of England. It is this feature of Scottish politics that lends credence to "uneven development" accounts of political differences between Scotland and England. But this "radical" vote is splintered – and has been for many years. In urban-industrial and mining areas, the Labour party has become the dominant if not always unchallenged political voice. Elsewhere this vote is shared by Labour, Liberals, and the SNP. In many places this vote, when totalled, far exceeds that of the Conservatives. The Liberal hegemony died in the 1920s. It has not been replaced.

CONTEXTS FOR ELECTORAL BEHAVIOR
How can these aggregate patterns of political behavior be explained in terms of patterns of social and economic life? Historically, Scotland has been geographically differentiated in various ways during the course of its modern social and economic history. The most outstanding feature of Scottish history since the 17th century has been the differentiated pattern of industrialization and associated pattern of social change. A major aspect of this has been a tremendous redistribution of population (Turnock 1982 : 40, 113–15). Central and southwest Scotland gained population during most of the period 1700–1960, while other sections suffered net losses, or only maintained their relative shares (Lenman 1981).

Not surprisingly, industrialization has been similarly concentrated. Economic growth in the 19th century was focused on west central Scotland. By 1911 this region contained 47.4 percent of Scots, when in 1841 it had accounted for only 34.5 percent. In this region were to be found all the heavy industries that characterized Victorian growth, from textiles and mining to shipbuilding and a full range of iron making and engineering activities.

Though the West dominated growth in the 19th century, it did not have a complete monopoly. Two other central regions, also generated high rates of growth (Lee 1983:8). The Lothian region around Edinburgh continuously augmented its share of population in the late 19th century. So too did the Central and Fife regions. Here again manufacturing in textiles, engineering, and mining were the major activities.

The differential rate of population growth and industrial concentration in these regions was necessarily at the expense of growth elsewhere. Each of the other Scottish regions, Highland, Grampian, Tayside, Borders, and Galloway, experienced *continuous* population decreases throughout the 19th century and well into the 20th century. The two southern regions had populations in 1971 similar in size to those of 1841. The Highland region suffered *absolute* population decline until 1961. Though Tayside and Grampian did enjoy some growth, concentrated mainly in Dundee and Aberdeen respectively, it was at a rate below that of the central regions (Lee 1983:9). These regions remained much more oriented towards agriculture than the others.

Since World War I, and more especially between 1960 and 1980, the traditional industries of Scotland have collapsed. At the eve of World War I, Clyde shipyards built not only one-third of British shipping tonnage but 18 percent of total world output (Slaven 1975:178). Such a large share of world production could not be sustained in the face of competition from industrializing countries and the downturn in world demand in the interwar period (Slaven 1975, Campbell 1980). At least in the interwar years, overcommitment to heavy industries cost Scotland, particularly the west, dearly. Scottish unemployment averaged 14 percent from 1923–30, against the UK's 11.4 percent. In 1931–8 this rose to 21.9 against 16.4.

World War II offered an economic respite. The shipyards were busy again. But the Scottish economy again failed to diversify. The new growth industries – motor vehicles, electrical goods, chemicals, artificial fibers – were not locating in Scotland. Indeed the industrial base of Scotland narrowed in the 1950s. As a percentage of total output the traditional "staples" had suffered a steep decline from 1907 until 1935, which then levelled off. By 1960 they still accounted for about one-third but by 1976 this had fallen to one-fifth (Table 7.1). More seriously, there was a rapid decline from the 1930s to the 1950s in "other" manufacturing industries – clothing, food and drink, paper, chemicals, timber – largely a result of English competition (Harvie 1981). Although expansion of the service sector compensated somewhat, the general trend was towards a narrowing of the industrial base.

In the 1950s most firms were largely Scots-owned and small-to-medium in size. Large factories were almost entirely a feature of the heavy industries

Table 7.1   Trends in major sectors of economic output in Scotland (percent).

	1907	1924	1935	1951	1960	1976
staples: agriculture, fisheries, mining, steel, engineering, and textiles	*53*	48	39	33	30	20
other manufactures	*26*	*27*	*27*	14	15	14
construction	*5*	*4*	*3*	5	6	9
service sector	*16*	*21*	*31*	48	49	57

Note: Italicized figures are estimates.
Source: Harvie (1981 : 38).

(iron and steel, engineering, shipbuilding). These large factories were concentrated in the central regions, more especially in or near Glasgow (Table 7.2). In the 1960s, however, new industries changed this picture, notably the expansion of the electronics industry from 7,500 jobs in 1959 to 30,000 in 1969. Many of these new jobs were located in the Scottish "new towns" – Glenrothes, Cumbernauld, East Kilbride, Livingston, Irvine – or outside the traditional industrial areas.

The new jobs were provided largely by foreign, particularly American firms (Firn 1975). As traditional heavy industries continued to decline they were displaced by the branch-plants of multinational firms. Many of these were keen to avoid "traditional" industrial areas with their skilled and highly unionized male labor forces. Female workers were both cheaper and more

Table 7.2   Factories in Scotland (1938).

A. *Location*	Large-scale (1,500 + employees)	Medium-scale (250–1,500 employees)
1 Scotland outside Central Lowlands	2	44
2 Central Lowlands outside Glasgow area	12	187
3 Glasgow area (Lanark, Renfrew, Dunbarton)	35	143

B. Scottish ownership of large-scale factories, by sector

	Total	Scots owned
engineering	12	7
textiles and derivatives	11	10
metallurgy	8	7
shipbuilding	7	7
co-operative works	3	3
miscellaneous	8	6

Source: Harvie (1981 : 39).

Table 7.3 The impact of North Sea oil in Scotland.

A. Employment in companies wholly related to the North Sea oil industry (thousands)

	Central and Lothian	Fife	Grampian	Highland	Strathclyde	Tayside	Islands	Scotland
1973	–	–	–	–	–	–	–	5.29
1974	1.07	1.21	4.81	4.86	1.17	0.27	0.09	13.47
1975	0.42	1.62	8.97	4.45	3.30	1.10	0.21	20.05
1976	0.69	2.03	11.54	6.78	4.23	1.44	0.40	27.10
1977	0.62	0.81	15.68	7.09	1.92	1.76	0.77	28.63
1978	0.55	1.38	22.89	6.00	0.50	2.05	0.63	33.99
1979	0.52	2.26	28.06	4.81	0.77	2.32	2.92	41.76
1980 Q1	0.74	0.88	32.92	4.61	2.51	2.07	3.07	46.80
Q2	0.86	0.81	32.32	4.35	2.73	1.81	3.46	46.34
Q3	0.77	0.82	32.64	4.76	2.99	1.82	4.02	47.82

B. Drilling activities in UK northern waters

	Wells drilled			Oil production, northern North Sea (million tonnes)
	Exploration	Appraisal	Production	
1967	7	–	–	–
1968	1	–	–	–
1969	8	–	–	–
1970	10	–	–	–
1971	17	4	–	–
1972	23	4	–	–
1973	34	13	–	–
1974	59	31	–	–
1975	75	35	8	1.1
1976	51	26	47	11.6
1977	58	27	89	37.3
1978	32	20	85	52.8
1979	13	6	53	76.1

Source: Lythe & Majmudar (1982 : 144).

Table 7.4   The regional structure of Scottish employment, 1977.

	Primary		Manufacturing		Public utilities		Services		Total number
	'000	% of total	'000	% of total	'000	% of total	'000	% of total	'000
Borders	4.2	10.9	14.2	37.0	3.1	8.1	16.9	44.0	38.4
Central	3.5	3.3	37.5	35.4	10.4	9.8	54.6	51.5	106.0
Dumfries Galloway	6.3	12.3	12.4	24.3	5.3	10.4	27.1	53.0	51.1
Fife	12.6	10.0	42.3	33.7	10.5	8.4	60.3	48.0	125.7
Grampian	14.7	8.2	38.3	21.5	19.0	10.6	106.5	59.7	178.5
Highland	4.2	5.8	11.6	15.9	11.3	15.5	46.0	62.9	73.1
Lothian	12.7	3.9	67.9	21.0	26.9	8.3	215.3	66.7	322.8
Strathclyde	17.5	1.8	342.4	34.5	87.2	8.8	545.6	55.0	992.7
Tayside	7.2	4.5	45.5	28.4	13.9	8.7	93.7	58.5	160.3
Orkney	0.8	13.6	0.5	8.5	0.8	13.6	3.8	64.4	5.9
Shetland	0.4	4.7	1.1	12.8	2.3	26.7	4.8	55.8	8.6
Western Isles	0.3	4.2	1.1	15.3	1.1	15.3	4.7	65.3	7.2
Scotland	8.5	4.1	6.5	29.7	19.2	9.3	117.9	56.9	2,071.0

Source: Scottish Abstract of Statistics, no. 10 (1981), from Lythe and Majmudar (1982 : 172).

docile, particularly in areas with no or a limited "industrial tradition". In some places, such as Linwood and Dundee, these new jobs came to dominate employment. Over the past ten years, however, some of these jobs have disappeared as the multinationals that created them have "retrenched" and reorganized their global operations (Hood & Young 1982).

The most recent "shock" to the Scottish economy has been the discovery and exploitation of oil from the North Sea. Since the early 1970s this has had a marked impact on the northeast of Scotland, particularly the Grampian region (Table 7.3). But elsewhere the economic and financial impact of the new oil industry has been limited (Chapman 1982:126, Lythe & Majmudar 1982:142–7).

The regional structure of the Scottish economy has undergone a series of transformations over the past 100 years. The West Central or Strathclyde region is still the major industrial region, but this regional bias is not as marked as 50 years ago (Table 7.4). The service sector is now the most important in all regions. But the regions are still economically distinctive from one another.

One important consequence of this differentiation is an uneven "welfare geography" in which some places and regions have higher incomes and lower levels of "social deprivation" than others. There is an extremely uneven geographical distribution of income in Scotland (Table 7.5). The highest incomes are found in Edinburgh (Midlothian), the lowest per capita incomes are found in Dunbartonshire, West Lothian, Argyll, and Bute – both urban and rural areas. The map of "social deprivation" is similar

Table 7.5 Total net income and income per capita, by region, 1959/60 and 1967/8.

	Net income, 1959/60			Net income, 1967/8		
	Total (£m)	Per capita (£)	Per capita income as % of Scottish average	Total (£m)	Per capita (£)	Per capita income as % of Scottish average
Aberdeen, Banff, Moray, Nairn	93.5	215.8	89	152.7	361.9	94
Angus and Kincardine	77.4	252.7	104	115.1	380.1	99
Argyll and Bute	14.2	200.8	83	19.3	271.6	71
Ayr	79.6	230.4	95	108.2	309.0	81
Berwick, East Lothian, etc.[a]	34.7	223.6	92	50.4	331.7	86
Caithness, Inverness, etc.[b]	39.6	180.8	74	76.7	351.2	92
Clackmannan and Kinross	11.6	238.2	98	16.5	334.0	87
Dumfries, Kircudbright and Wigtown	34.1	226.4	93	49.1	340.5	89
Dunbarton	40.2	220.4	91	51.5	235.2	61
Fife	71.2	218.6	90	97.5	301.5	79
Lanark	403.0	246.8	101	683.8	439.0	114
Midlothian	167.5	288.2	119	282.4	477.2	124
Perth	29.9	237.5	98	39.0	312.8	82
Renfrew	99.9	295.2	121	153.7	430.2	112
Stirling	46.2	236.7	97	64.4	321.1	84
West Lothian	14.8	156.8	64	29.2	280.5	73
Total Scotland	1,257.4	243.2	100	1,989.5	383.5	100

[a] Includes Peebles, Roxburgh, and Selkirk.
[b] Includes Ross and Cromarty, Orkney, Sutherland, and Zetland.
Sources: Calculated from Inland Revenue Reports: Annual Reports of the Registrar General (from Johnston *et al.* 1971 : 143).

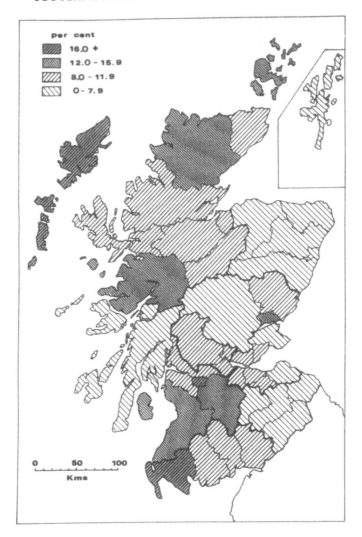

Figure 7.2 The provision of free school meals in Scotland by district, 1976 (administrative subdivisions in Strathclyde).
Source: Knox & Cottam (1981 : 167).

(Fig. 7.2). Using the provision of free school meals as an indicator of poverty, it is clear that the incidence of poverty is greatest in the two extremes of the Scottish "socioeconomic environment": Glasgow and the Western Isles. Location quotients, which control for differences in population size, show that deprivation is localized in seven districts – four rural and three urban (Table 7.6).

Deprivation is a multidimensional concept and, as Knox (1979) has pointed out, cannot be reduced to a unitary measure. An analysis of nine key

Table 7.6   The localization of deprivation.

District	Location quotient
Western Isles	2.27
Glasgow	1.68
Wigtown	1.30
Ayr	1.15
Lanark	1.12
Dundee	1.07
Edinburgh	1.08

Source: Knox & Cottam (1981 : 166).

variables by Knox and Cottam (1981), covering unemployment, depopulation, age dependency, housing conditions, car ownership, infant mortality, and free school meals, reveals however that Figure 7.2 is not far off the mark.

There are also important cultural and social–institutional differences between parts of Scotland that cross-cut and reinforce the basic material and welfare cleavages. A relatively minor one is language. Gaelic-speakers are a clear majority only in the Western Isles. Elsewhere they are a small and declining minority: Gaelic speaking in Scotland as a whole has declined from 11 percent in 1891 to 3 percent in 1961. A much more significant one is religion. Religion is often viewed as a relict feature in the landscape of secularizing industrialization. But in Scotland, as in Poland, Iran, and North America, religion continues to help define both communal and class boundaries simply because churches have become associated with minority rights and class interests (Turner 1981). The established Church of Scotland has helped to define a "Scottish" cultural identity that has been particularly appealing to the value and interests of the lower middle-class and "respectable" working-class in small- and medium-size towns (Turner 1981 : 90). But there are major denominational divisions within Scotland that challenge and even contradict the claim of the Church of Scotland to be *the* national church in anything other than name. These merit close attention.

The major religious division in Scotland is that of Protestant and Catholic. Apart from some indigenous Catholic communities in the Western Isles and in Inverness, the bulk of the Catholic population in Scotland is descended from Irish immigrants. Thus the distribution of Scottish Catholics reflects the pattern of Irish immigration. Most immigrants were attracted to the industrial areas, and it is in those areas, particularly in the West Central or Strathclyde region, that most Scottish Catholics are still to be found (Table 7.7). There is some evidence of a spreading out, but what is remarkable is how *limited* this has been (Table 7.8). The Catholic population has also grown in size, to the extent that there is today a much smaller gap between the number of Catholics and the number of practicing members of the

Table 7.7 The Catholic population of Scotland, 1977.

REGION/DISTRICT (Only districts with populations exceeding 100,000 included plus Argyll and Bute)	Estimated Catholic population		Baptisms		Marriages	
	Number	% of population	Number	% of live births	Number	% of all marriages
Highland region	7,500	3.9	227	8.8	75	5.6
Orkney Islands area	100	0.6	8	3.5	1	0.9
Shetland Islands area	100	0.5	5	1.5	–	–
Western Islands area	4,000	13.6	59	15.5	14	11.1
Grampian Region	7,900	1.7	213	3.8	125	3.8
Aberdeen City	4,100	2.0	101	4.8	62	3.6
Tayside region	43,000	10.7	727	15.9	269	9.7
Dundee City	34,000	17.6	573	26.2	195	13.8
Perth and Kinross	6,600	5.6	92	7.3	53	6.9
Fife region	21,300	6.3	425	10.4	174	6.7
Kirkcaldy	9,100	6.1	181	9.7	55	4.7
Dunfermline	10,300	8.3	193	12.1	92	8.9
Lothian region	75,500	10.1	1,216	14.4	498	9.2
Edinburgh City	42,000	9.1	649	14.0	307	8.5
West Lothian	21,900	17.1	345	18.2	112	14.0
Borders region	3,900	3.9	77	7.1	28	4.6
Central region	26,900	9.9	462	14.4	183	10.0
Falkirk	15,300	10.6	257	14.8	101	10.3
Strathclyde region	624,000	25.3	9,462	31.4	4,089	22.7
Argyll and Bute	4,300	6.7	121	14.5	35	8.7
Renfrew	47,700	22.6	718	28.2	330	22.6
Inverclyde	37,300	36.1	599	47.8	267	31.9
Glasgow City	254,800	30.6	3,620	38.4	1,731	24.9
Monklands	46,500	41.8	728	50.9	317	34.4
Motherwell	51,100	33.2	717	36.0	326	28.0
Hamilton	29,200	27.1	502	33.1	162	27.3
Cunninghame	19,100	14.1	335	10.9	140	15.1
Kyle and Carrick	9,100	8.0	150	11.9	67	8.2
Dumfries and Galloway region	8,300	5.8	149	9.5	62	5.9
Scotland	823,500	15.8	13,030	20.6	5,518	14.8

Source: Darragh (1978 : 240). Region/district populations, live births, Catholic marriages and percentages from *Annual Report* for 1977 of Registrar General for Scotland. Catholic population and baptisms from *Catholic Directory for Scotland*, editions for 1978 and 1979.

Table 7.8 Catholic population estimates, by county, 1851–1971.

	1851		1951		1971	
County	Number	% of population	Number	% of population	Number	% of population
Aberdeen	4,700	2.2	7,300	2.4	5,200	1.6
Angus	4,400	2.3	37,500	13.6	39,300	14.1
Argyll	2,500	2.8	2,700	4.3	3,300	5.5
Ayr	7,700	4.1	29,900	9.3	35,500	9.8
Banff	6,200	11.4	2,500	4.9	1,500	3.4
Berwick	–	–	600	2.3	400	1.9
Bute	400	2.4	1,300	6.9	1,100	8.3
Caithness & Sutherland	–	–	100	–	600	1.4
Clackmannan & Kinross	–	–	2,700	6.0	4,100	7.9
Dumfries	4,500	5.8	4,700	5.5	5,700	6.5
Dunbarton	1,400	3.1	36,300	22.1	59,100	24.9
East Lothian	–	–	2,400	4.6	4,400	7.9
Fife	1,400	–	18,700	6.1	20,900	6.4
Inverness	10,800	11.2	10,200	12.0	9,400	10.5
Kincardine	–	–	100	–	200	–
Kirkcudbright	1,900	4.4	1,200	3.8	1,200	4.3
Lanark	66,300	12.5	415,500	25.7	431,700	28.3
Midlothian	10,600	4.1	59,100	10.4	53,600	9.0
Moray	1,100	2.8	1,000	2.0	900	1.7
Nairn	–	–	200	2.2	200	1.8
Orkney & Shetland	–	–	100	–	100	–
Peebles	300	2.8	500	3.0	600	4.4
Perth	2,000	1.4	5,600	4.4	7,100	5.6
Renfrew	12,700	7.9	71,300	22.0	88,800	24.5
Ross & Cromarty	–	–	500	–	500	–
Roxburgh	600	1.2	1,300	2.8	1,400	3.3
Selkirk	–	–	900	4.2	1,500	7.2
Stirling	2,700	3.1	21,900	11.7	25,900	12.4
West Lothian	–	–	13,400	15.1	17,600	16.2
Wigtown	3,700	8.5	1,100	3.4	1,100	4.0
Scotland	145,600	5.1	750,600	14.7	822,900	15.7

Source: Darragh (1978 : 228).

Church of Scotland (Darragh 1978 : 225). However, the Catholic birth rate has diminished considerably in recent years, to a level only slightly above that of Scotland as a whole (Darragh 1978 : 231). So the growth in numbers is probably slowing down.

Outside of the Strathclyde region the rate of "mixed marriages" is fairly high, so the social and political *meaning* of Catholicism is probably different there (Darragh 1978 : 237). The rate of mixed marriages measures both social insularity and the power of the clergy. The Catholic clergy in Scotland have had the reputation of discouraging mixed marriages more actively than any other (Gallagher 1981:26). In Strathclyde there has been a resistance to assimilation. On the south side of Glasgow, in northern Lanark, and on Clydeside, Irish communities emerged that retained a separate identity long after the descendants of the original immigrants had adopted the speech and folkways of west central Scotland. Sectarian tensions were high in the depression years of the 1930s, especially in places such as Coatbridge, Greenock, Port Glasgow, and Govan (Glasgow), which also had large Irish *Protestant* populations (Gallagher 1984). As recently as 1976, the militantly Protestant Orange Order claimed 80,000 members in Scotland (Wilson 1976). So, even though intergroup tensions have declined since the 1930s, there is a persisting identification with religious labels that, at least in Strathclyde, continues to be of potential social and political importance. This also colors the "image" of this region elsewhere in Scotland (Mackenzie 1981).

Major divisions occurred within the Church of Scotland in the 19th century. The most important one was a fracture along the Highlands–Lowlands divide. The Highlands had been proselytized for Presbyterianism after the Jacobite rebellion of 1745. What is clear, however, is that although the Church of Scotland had won control of the Highlands from Catholicism and Episcopalianism by the end of the 18th century, "there was little sign of popular enthusiasm for, or attachment to, the Establishment" (Hunter 1976:94). Presbyterianism only took strong hold through religious revivalism and the distribution of Gaelic bibles in the early 19th century. During the Disruption of 1843, the bulk of the ministers and laity in the north of Scotland went over to the Free Church. Though the base of the Free Church lay in the Lowlands, it was in the Highlands that it became a medium for religious and social separatism.

It is apparent, then, that both material and religious–cultural cleavages take a marked geographical form, both now and in the Scottish past. These have had two major political consequences. One has been the containment of Labour, and the other has been a geographical limitation on the appeal of the SNP.

Not surprisingly, it was in the more industrialized parts of Scotland, particularly Strathclyde, that a more or less unequivocal class-based politics took root in the late 19th century. This has been reinforced by the overwhelming support of Scottish Catholics for the Labour party. An estimated 79.3 percent of Scottish Catholics voted for the Labour party in the February 1974 general election, a time when it was under strong pressure from the SNP (Clarke & Drucker 1975:12). Budge and Urwin (1966) estimated that in the 1960s the Catholic vote for Labour was more solid than that of the Strathclyde working class as a whole. This commitment to the Labour party extends back to a collective choice (stimulated by the clergy!)

made at the time when Irish Home Rule was finally achieved. From one point of view this was a defensive act. The Conservative party was the party of the Protestant and Unionist establishment. Alternatives to the Labour party in 1918–22 were "revolutionary" parties that took their atheism too seriously for them to receive any clerical or much popular support. From another point of view the Labour–Catholic alliance was forged in the period of the Education (Scotland) Act, 1918, because the Labour party was less hostile to church schools (Turner 1981 : 104).

However, at the same time the so-called "Irish issue," the presence of Irish Catholic immigrants in Protestant Scotland and their presumed disloyalty to the British state, has declined in importance in Strathclyde (Gallagher 1981). The increased dominance of the Labour party in Strathclyde since World War II and the decline of the Conservative vote in the region certainly point towards this (Miller 1981, Bodman 1985). The SNP has had very little success among the region's Catholics (Brand & McCrone 1975:417, Drucker & Brown 1980:50). They are suspicious of how they fit within the overtly Scottish (and thus Presbyterian) ideology of Scottish nationalism, at least as embodied in the SNP. The dramatic increase in SNP votes in 1974 and the equally dramatic decline in 1979 suggest a temporary defection of presumably non-Catholic voters engaged in a 'protest' vote rather than a realignment of support for political parties (Miller et al. 1977).

Yet too much can be made of the Labour–Catholic connection in Strathclyde. Only one-third of Labour's support there comes from Catholics (Rose 1971 : 168). Partly this is an occupational class effect; working-class people in Strathclyde, whatever their religion, prefer Labour. There are more "working-class" constituencies and fewer "middle-class" constituencies in Strathclyde than in most other parts of Britain (Kellas & Fotheringham 1976 : 154). But, also, as the Irish immigrant population has ceased to be a "low-wage" threat and new housing developments have undermined patterns of religious residential segregation, significant barriers to class formation have broken down. The working-class "potential" or capacity of the places in this region has thus mobilized increasingly on behalf of the Labour party rather than other parties with a less favorable working-class image.

But the shift to Labour also reflects a "consumption sector" cleavage in housing (Dunleavy 1979). Many more people in urban-industrial Scotland live in council housing than in owner-occupied housing; 51.8 percent of dwellings in Strathclyde are in the public sector (Regional Trends 1981 : 30–1). This compares to a "high" of 58.5 percent in the Central region and a low of 15.1 percent in Fife. The Scottish average is 45.7 percent. There is evidence that this affects political behavior. For example, a comparison of two Glasgow constituencies, one Labour-held and the other Conservative-held in 1970, reveals that their major difference lies in housing composition (Table 7.9). Council-housing and owner-occupied housing are also

Table 7.9 Housing provision, occupation and political behavior, Glasgow: Cathcart and Craigton, 1970.*

	Occupation		Type of housing	
	Non-manual (%)	Manufacturing (%)	Council housing (%)	Owner-occupied (%)
Cathcart (Conservative)	39.3	52.4	41.1	33.3
Craigton (Labour)	37	63.1	64.4	18.9

* Figures do not sum to 100% because of other occupations and other types of housing.
Source: Kellas & Fotheringham (1976 : 158).

increasingly segregated (Kellas & Fotheringham 1976 : 147). And this polarization is compounded by the new provision of comprehensive education along territorial lines, which limits the movement of working-class children into secondary schools in which they can mix with middle-class children (see also Agnew 1978, 1981c). Housing provision, therefore, is hardening pre-existing class divisions in a number of ways as well as providing a separate "consumption cleavage."

More generally, rather than resting just on housing consumption preferences and a dependence upon Labour housing programs, support for the Labour party in Strathclyde rests upon a well-established tradition of "municipal socialism." In the absence of private action, the severity of economic, housing, and environmental problems in west central Scotland, especially Glasgow, has stimulated public intervention. It is on local politics and associated local concerns that the Labour party in the west of Scotland has built its popular base. The decline of the heavily unionized traditional industries has not led to the demise of local working-class consciousness. If anything, the opposite has happened. A sense of common victimization and collective self-interest has developed a stronger identity with the Labour party among *all* working-class elements than was the case in the heyday of "Red Clydeside" (1918–22), or when the large factories were working at full steam (Miller 1981). This is why there has been a steady swing to Labour in Strathclyde over the past 40 years (Bodman 1983). Of course, there has been a parallel swing away from the "revolutionary" politics and tactics of earlier times as the Labour party has "co-opted" the local population for parliamentary politics (J. Smith 1984).

Before the destruction of the distinctive pre-industrial social order in the Highlands by a combination of military suppression and economic "clearances" (population removals), Scotland was indeed "two nations" separated by language, family structures, economic organization, and religion. With "integration," however, a new distinctiveness evolved. In the 19th century, prior to changes in landholding and at the same time as tremendous depopulation was occurring, the Highlands was the scene of widespread social and political discontent (Dunbabin 1974:263–71, Hunter 1974). Rural discontent took the form mainly of land invasions and other so-called "agrarian outrages." The poverty of the region and landlord policies of clearance and eviction also produced a more organized form of protest: the Crofters' party (Hanham 1969a, Hunter 1974, 1975). Hunter (1975:187) argues that the "highland land war" of the 1880s was analogous "both in its origins and in the aims of the tenants involved in it, to the Irish land war of the same decade." The Highland Land League or Crofters' party was created in 1881 to address questions specific to northern Scotland. It was not a nationalist party. But it did support Irish Home Rule, and had close ties to the Scottish Home Rule Association and the "Gaelic Lobby" (Hunter 1975).

The Crofters' party won five parliamentary seats in the 1885–6 elections. The government took the hint, and a series of Acts were passed establishing peasant proprietorship. By 1892 the Crofters' party had merged completely with the Liberals (Hanham 1959). The passage from activism to passivity was dramatic. It reflected both readiness to compromise, and the emergence of Irish Home Rule as a divisive issue. In each case the hold of "independent" Presbyterianism was crucial. The Free Church, and after 1893 its splinter, the Free Presbyterian Church, though anti-landlord, provided a powerful connection between "the incipient radicalism of the crofting population and the mainstream of Scottish Liberal and radical politics" (Hunter 1976:105). They also, by means of their fatalistic theology and Calvinist austerity, suppressed the established folk culture – folklore, traditional music, and other customs. Gaelic remained as a secondary church and folk language, but one that was residual rather than dynamic. Thus, though a vehicle for rural radicalism in the 1880s, the Free Church became a barrier to continuing radicalization of either a nationalist or socialist variety.

In this century, depopulation has continued apace in the Highlands. Crofting, small-scale farming with security of tenure or owner–occupancy and rights to pasture on common grazing, has declined in all but a few places (Carter 1974, Gillanders 1968). This is partly the result of an aging population and partly the result of an in-migration of lowland Scots and others taking over crofts as weekend, summer, or permanent homes. The distinctiveness of the Highlands now rests on an odd amalgam of indigenous and declining Gaelic Calvinism and exogenous English Hedonism. This has produced an extremely volatile brand of politics (Table 7.10).

Since 1945 the fortunes of all parties, in terms of votes cast and seats won, have fluctuated widely. Some national issues (British and Scottish) do appear to excite controversy – for example, the question of British entry into the EEC raised both material and ideological objections – but by and large, local

Table 7.10 Election results in the Highlands, 1945–79.

Constituency	Party	1945	1950	1951	1955	1959	1964	1966	1970	Feb. 1974	Oct. 1974	1979
Argyll	SNP	–	1.7	–	–	–	–	–	29.9	48.8*	49.7*	31.9
	Cons.	56.5*	66.5*	68.1*	67.6*	58.4*	47.2*	43.2*	48.8*	38.6	36.6	36.2*
	Lab.	31.9	31.8	31.9	32.4	25.9	28.9	30.1	25.3	12.6	13.6	15.9
	Lib.	11.6	–	–	–	15.7	23.0	16.7	–	–	–	15.4
Caithness and Sutherland	SNP	–	–	–	–	–	–	–	15.5	16.1	23.9	27.9
	Cons.	–	35.9*	49.3*	56.5*	–	33.6	22.0	22.4	21.5	18.8	30.4
	Lab.	33.4	29.7	34.1	29.0	34.6	30.3	39.1*	36.7*	36.1*	35.3*	41.0*
	Lib.	33.1	34.5	16.6	14.5	–	36.1*	38.9	25.4	26.2	22.0	–
	Ind. Cons.	33.5*	–	–	–	65.4*	–	–	–	–	–	–
Inverness	SNP	–	–	–	–	–	–	–	7.1	16.6	29.6	20.6
	Cons.	–	45.5*	64.5*	44.4*	44.3*	33.9	32.9	31.5	26.7	22.0	24.8
	Lab.	34.6	31.8	35.5	22.7	22.8	26.3	27.7	23.0	17.9	15.6	20.5
	Lib.	22.2	22.7	–	32.9	32.9	39.8*	39.4*	38.4*	38.7*	32.3*	33.7*
	Ind. Lib.	43.2*	–	–	–	–	–	–	–	–	–	–
Ross and Cromarty	SNP	–	–	–	–	–	–	–	11.7	23.0	35.7	23.5
	Cons.	–	62.6*	64.2*	62.3*	47.2*	32.1	27.6	33.2*	36.1*	38.9*	42.4*
	Lab.	37.2	37.4	35.8	37.7	29.1	27.7	30.4	26.0	19.8	16.8	20.2
	Lib.	–	–	–	–	23.7	40.2*	42.0*	29.1	21.1	8.5	13.5
	Ind. Lib.	62.8*	–	–	–	–	–	–	–	–	–	–
Western Isles	SNP	–	–	5.0	–	–	14.0	–	43.1*	67.0*	61.4*	52.5*
	Cons.	21.3	–	40.7	42.7	46.4	–	20.2	18.5	6.9	8.3	10.5
	Lab.	45.7*	53.2*	48.8*	57.3*	53.6*	55.1*	61.0*	38.4	25.9	24.7	32.1
	Lib.	33.0	44.2	5.6	–	–	30.9	18.8	–	–	–	4.6
	Ind. Nat.	–	2.7	–	–	–	–	–	–	–	–	–

* *Winners.*

Sources: Craig (1969), Kellas (1979), Mitchell & Boehm (1966), *The Times* (12 Oct. 1974), *Daily Express* (5 May 1979).

social and economic concerns are central (Kellas 1979:214–15). This is especially the case in the Western Isles, the most homogeneously Gaelic-Calvinist constituency, where high levels of deprivation have led voters towards a consistent support for the most "radical" parties of the day: Labour and SNP. What is often more important, however, is that a candidate be a strong constituency advocate rather than a "party man." In peripheral areas an MP links the localities with the centers of government, and must move easily between them (Barth 1963). In this respect, local people with local contacts and interests are preferred to outsiders (Kellas 1979:214–15). In 1983, to use the latest example of volatility, three Liberal/SDP Alliance candidates, two of whom were SDP candidates, were elected out of five seats contested. One continuity is unquestionable: the Highlands is a region in which attachments to the two major British parties have been and remain weak. In 1983 the Alliance (40.7 percent) and SNP (16.7 percent) drew 57.4 percent of the vote in the Highlands. The Labour party acquired a mere 15.7 percent and the Conservatives a meager 26.7 percent.

The interweaving of place and politics in the rest of Scotland is considerably more complex and fragmented than in Strathclyde and the Highlands. But there are several regions in which local conditions and changing economic relationships with the outside world have brought about a shift towards nationalism and the SNP. The most notable are the Grampian, Tayside, and Galloway regions, and parts of the Central region.

If the regional pattern of seats contested is compared to the pattern of seats won in the period 1970–83, it is clear that three regions – Highlands, Grampian, and Tayside – stand out as regions of above-average performance for the SNP (see Table 7.11). This bias is also reflected in the regional distribution of high shares of votes cast by constituency; and even when a control for seats contested is introduced, the regional pattern persists (Table 7.12). Strathclyde, Lothian, and Borders are regions in which support for the SNP is particularly low and "soft" compared to other regions.

Disaggregation reinforces the evidence for the place-specific pattern of support for the SNP. Northern Grampian, Eastern Tayside, Galloway, and Western Isles are consistently the most nationalist areas. In 1970 the following constituencies had the highest SNP vote shares (20 percent or more): Aberdeenshire East, Angus South, Argyll, Banffshire, Western Isles, Galloway, Hamilton, Kinross and West Perthshire, West Stirlingshire, and West Lothian. In 1979 the following occupied a similar position: Aberdeenshire East, Angus South, Argyll, Banffshire, Dundee East, Galloway, Moray and Nairnshire, Perth and East Perthshire, Clackmannan and East Stirlingshire, and Western Isles. Of the ten high shares in 1970, six reappear in 1979; and of the four others, two are in the same regions (Central and Tayside) in both 1970 and 1979. The 1983 election involved new constituency boundaries, but though this disrupted links between places and parties, a similar pattern to that of 1979 is evident. The following nine constituencies

Table 7.11 The SNP and general elections since 1929.

Date	Highlands (incl. Argyll)	Strathclyde	Central	Lothian	Fife	Tayside	Grampian (incl. Orkney and Shetland)	Borders	Dumfries and Galloway
A. Seats contested									
1929	—	2	—	—	—	—	—	—	—
1931	1	3	—	1	—	—	—	—	—
1935	2	5	—	—	—	—	—	—	—
1945	—	3	1	1	1	2	1	—	—
1950	1	1	—	—	—	1	—	—	—
1951	—	—	1	—	—	1	—	—	—
1955	—	1	1	—	—	1	—	—	—
1959	—	4	2	1	2	2	1	—	1
1964	—	9	3	3	4	2	2	1	1
1966	—	29	3	8	4	2	1	—	1
1970	5	33	3	9	4	6	6	2	2
1974 (Feb.)	5	33	3	9	4	6	6	2	2
1974 (Oct.)	5	33	3	9	4	6	7	2	2
1979	5	33	3	9	4	6	7	2	2
1983	5	32	4	10	5	5	7	2	2
B. Seats Won									
1970	1	0	0	0	0	0	0	0	0
1974 (Feb.)	2	0	1	0	0	1	3	0	0
1974 (Oct.)	2	1	1	0	0	3	3	0	1
1979	1	0	0	0	0	1	0	0	0
1983	1	0	0	0	0	1	0	0	0

Source: Author's calculations based on F. W. S. Craig, *Minor Parties at British Parliamentary Elections 1885–1974* (London: Macmillan, 1975); "How the Nation Voted," *Daily Telegraph*, 5 May 1979; "Members of the New House of Commons," *The Times*, 11 June 1983.

Table 7.12 Regional distribution of high levels of support for the SNP (over 20% of vote).

Election date	Highlands (incl. Argyll)	Strathclyde	Central	Lothian	Fife	Tayside	Grampian (incl. Orkney and Shetland)	Borders	Dumfries and Galloway
A. Number of candidates receiving high shares									
1964	–	–	–	1	–	1	–	–	–
1966	–	–	2	1	–	–	–	–	–
1970	2	1	1	1	–	2	2	0	1
1974 (Feb.)	3	11	3	3	1	5	4	0	1
1974 (Oct.)	5	29	3	9	4	5	6	0	1
1979	2	0	1	0	0	4	3	0	1
1983	2	0	0	0	0	4	2	0	1
Total number of seats									
(1970–1979)	5	33	3	9	4	5	7	2	2
(1983)	5	32	4	10	5	5	7	2	2
B. Seats having high shares as percentage of seats contested									
1964	*	0	0	100	0	0	0	0	0
1966	*	0	66	33	0	20	0	*	0
1970	40	3.4	33.3	12.5	0	40	33.3	0	50
1974 (Feb.)	60	33.3	100	33.3	25	100	57.2	0	50
1974 (Oct.)	100	87.8	100	100	100	100	85.8	0	50
1979	40	0	33.3	0	0	80	82.9	0	50
1983	40	0	0	0	0	80	28.6	0	50

* = No contest.
Source: Author's calculations.

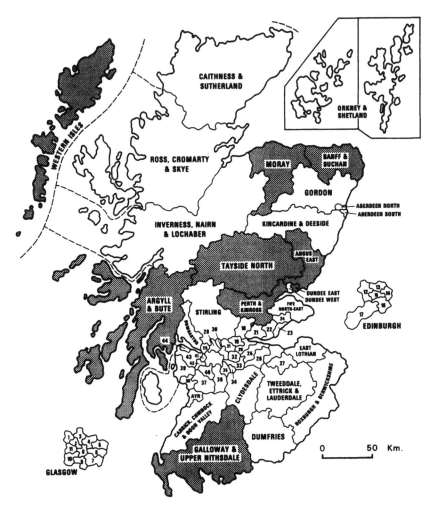

Figure 7.3 The 1983 Scottish constituencies. Constituencies where 20 percent or more of the vote was for the SNP in the general election of 9 June 1983 are shaded.

had SNP votes in excess of 20 percent (percentage of vote in parentheses): Western Isles (54.5), Dundee East (43.8), Banff and Buchan (37.4), Angus East (36.0), Moray (35.2), Galloway (30.8), Perth and Kinross (25.1), Argyll (24.6), and Tayside North (24.3) (Fig. 7.3). Of the nine first or second places for SNP in 1983, *seven* were in Tayside, Grampian, or Galloway.

Although often more local in orientation and less partisan in nature than parliamentary elections, the results of regional and district elections in the 1970s and early 1980s also suggest a regional bias in SNP support. For example, in regional elections the SNP has done much better in the Central region than elsewhere (Tables 7.13, 7.14, 7.15). The SNP has done

Table 7.13 Seats won in regional elections, 1974.

	Conservative	Labour	Liberal	SNP	Communist	Ind.	Other
Highland	2	4	3	1	1	37	–
Grampian	28	13	2	1	–	10	–
Tayside	22	15	–	–	–	9	–
Fife	10	26	–	–	1	3	2
Lothian	19	24	1	3	–	1	1
Central	4	17	–	9	–	3	1
Borders	7	–	3	–	–	13	–
Strathclyde	20	71	2	5	–	5	–
Dumfries & Galloway	–	2	–	–	–	33	–

Source: Bochel & Denver (1975 : 146).

Table 7.14 Seats won in regional elections, 1978.

	Conservative	Labour	Liberal	SNP	Communist	Ind.	Other
Highland	1	3	2	2	–	37	–
Grampian	33	13	–	2	–	4	1
Tayside	25	15	–	–	–	6	–
Fife	14	24	–	1	1	2	–
Lothian	18	26	1	3	–	1	–
Central	6	18	–	6	–	2	2
Borders	9	–	1	1	–	11	1
Strathclyde	25	72	2	2	1	–	1
Dumfries & Galloway	5	3	–	1	–	26	–

Source: Bochel & Denver (1978 : 62).

Table 7.15 Seats won in regional elections, 1982.

	Conservative	Labour	Liberal	SNP	SDP	Communist	Ind.	Other
Highland	1	5	2	2	–	–	42	–
Grampian	28	15	6	3	–	–	2	–
Tayside	27	12	–	5	–	–	2	–
Fife	10	27	2	1	2	1	2	–
Lothian	22	22	1	1	2	–	1	–
Central	4	22	1	5	–	–	2	–
Borders	8	–	3	–	–	–	12	–
Strathclyde	51	79	4	3	–	–	2	–
Dumfries & Galloway	4	4	2	3	–	–	22	–

Source: Bochel & Denver (1982 : 59).

particularly badly in Strathclyde regional elections. Many of the places in which the SNP now does well were once Liberal strongholds (Pelling 1967:386–97). In general they resisted the trend to Labour in Scotland as a whole and Strathclyde in particular after the 1920s. From the 1940s until the early 1970s they were areas of strength for the Conservatives. Ex-Liberal voters either moved to the Conservatives, abstained, or split between Liberal and Labour. In the 1970s, however, the non-Conservative segment of the population was again provided with a vehicle for expressing its concerns: the SNP. But what are these concerns?

In general terms, there have been two broad types of support for the SNP. One has come from farming and fishing interests opposed to British government and EEC farming and fishing policies. They have found allies in such groups as distillery workers and small-town professionals attached to "traditional" (i.e., 19th century) Scottish values and institutions. Continuing attachment to the Church of Scotland's vision of Scottish distinctiveness has been particularly important (Turner 1981:104–6). But there is also a system of agricultural organization in these places, particularly in the Grampian region, that has encouraged an antilandlord and antideferential attitude amongst farm workers and small farmers, but which is also strongly parochial (Kellas 1976:130, Carter 1976). Average farm size is small (Russell 1970). Profit margins are small. Many farms have faced increasing competition from larger units elsewhere, especially in beef production where the northeast was once preeminent (Bowler 1975). But farmers are still conscious of being "owners" and farm workers have decreased in number with increasing mechanization (Wagstaff 1972). There is little potential, then, for a "class" response to uncertainty and difficult times in such areas.

Similarly in fishing constituencies, the success of SNP candidates rests on "petit bourgeois" support for SNP policies. But in this case support is allied to small-town nostalgia for a Scotland that was historically remote even before bureaucracy intruded into local life. In many of these places, populations previously depoliticized or indifferent to party politics but hostile to state and union intervention have found in the SNP, usually represented locally by many elders and ministers of the churches as well as by local professionals, leadership in their struggles over industrial location policy, government reorganization, and development grants (Turner 1981:105).

The second type of support comes from people who have been mobilized by changing conditions rather than by the need to protect established ones. In particular, as industry and employment opportunities have spread throughout Scotland, they have shifted away from the large conurbations towards smaller cities and more rural locations (Dicken 1982). What is more, many of these jobs are nonunionized and with multinational firms (Scott & Hughes 1976, Hood & Young 1982). This is especially true of new jobs in the

Table 7.16 Location of ultimate ownership of Scottish manufacturing employment, by planning region, 1973 (percent).

Location of ultimate ownership	Glasgow	Edinburgh	Tayside	Northeast	Falkirk Stirling	Borders	Southwest	Highlands & Islands	All Scotland
Scotland	36.2	48.1	55.3	51.1	37.6	39.1	37.6	55.4	41.2
England	42.5	39.6	18.6	38.9	42.0	53.7	53.3	19.7	39.8
Rest of UK	–	1.0	–	–	–	–	–	0.7	–
EEC	2.0	–	0.8	1.8	2.1	2.6	–	–	1.5
Other Europe	0.7	–	–	–	0.8	–	1.9	1.0	0.6
North America	16.7	10.5	24.2	7.0	9.9	4.6	7.2	17.5	14.9
Other	–	–	0.3	–	–	–	–	–	0.1
Joint-owned	2.0	0.7	0.8	1.2	7.6	–	–	5.7	1.9
Total	100.0	100.0	100.0	100.0	100.0	100.0	100.0	100.0	100.0
Total numbers	338,510	98,485	49,290	33,300	34,885	15,865	7,485	10,880	590,700

Source: Firm (1975 : 408)

Tayside region, particularly around Dundee (Firn 1975:408) (Table 7.16). One political effect of this trend, particularly for migrants, may have been a disengagement from conventional political identification. This has created "political space" for the SNP, as illustrated by the success of the party in many of Scotland's new towns, especially Cumbernauld (McDonald 1975). But a reinforcing effect may have been the growth of a sense that Scots have lost control of their economy and are vulnerable to rapid disinvestment. The image of Scotland as a multinational branch-plant economy manipulated from New York may be particularly powerful among the employees of the multinationals. It is interesting to speculate that one cause of the SNP's failure in the Borders region is the lack of non-British employers there (Firn 1975:408).

Of course, one well-publicized correlate of growth in support for the SNP has been the discovery and exploitation of Scotland's North Sea oil. It is in the Grampian region, but in scattered places elsewhere too, that the discovery of oil has had its most profound and disruptive effects. The appeal "It's Scotland's Oil" is particularly effective in those districts where the oil industry has become well established since 1970 but from which the positive "spillover" effects have been less than anticipated by local populations (Moore 1982).

The SNP appeals to these "new" constituencies by stressing an image as the party of practical policies designed to fix the catastrophes brought about by central government (London) rule. This conflicts with its traditional ideological appeal to Scottishness and the practical interests of the "primary" constituency, and opens the SNP to the charge that it is merely a Scottish "third force" akin to the Liberals in England. But of course until the SNP downgrades its Presbyterian and romantic concept of Scottishness it is not likely to achieve success in the industrial core of Scotland with its lion share of Scotland's votes. Since the elections of 1974, and with the exception of Dundee, the SNP has faded into its rural, agricultural, and Presbyterian heartland. The "breakthrough" of the SNP into the new towns in 1974 was reversed in 1979. The traditional divisions between rural and urban, Lowland and Highland, Protestant and Catholic have survived SNP efforts to mobilize support around a popular party unencumbered by denominational, class, and regional interests (Turner 1981).

So it is in the small towns and farming communities of the Northeast, Tayside, and Galloway that the ideological appeal of Scottish nationalism in the form of the SNP is at its greatest. Elsewhere its "cultural consumption" side is secondary to its instrumental "investment" side, and thus can be undercut by class appeals. The consumption side of nationalism involves the so-called psychic rewards that the act of supporting nationalism provides at the time of acting (identification with a community, patriotism, affirmation of identity, racial prejudice, etc.). The investment side refers to "the expected benefits in material and symbolic status to be had if the party is

Table 7.17 Distribution of votes and seats for parties by regions, 1983 general election.

	Labour		Conservatives		Alliance		SNP	
	%	Seats	%	Seats	%	Seats	%	Seats
Borders	7.4	0	34.4	0	54.2	2	3.8	0
Central	41.2	3	25.5	1	20.1	0	12.7	0
Dumfries & Galloway	16.2	0	44.6	2	18.6	0	20.4	0
Fife	35.5	4	28.5	1	26.8	0	8.1	0
Grampian	18.8	1	37.7	4	26.5	1	16.6	0
Highlands	15.7	0	26.7	1	40.7	3	16.7	1
Lothian	35.6	6	30.3	4	26.3	0	7.4	0
Strathclyde	44.1	26	24.2	5	22.4	1	8.9	0
Tayside	20.8	1	33.6	3	15.8	0	29.3	1

Source: Author's calculation from *The Times*, 10 June 1983.

successful in its nationalist programme" (Green 1982:249). The Scottish-ness of the SNP provides a consumption rather than an investment appeal. SNP voters in the "nationalist regions," then, are not just Liberals in search of a party, as some would have it (Miller 1981:206). More to the point, the Liberals were all they used to have. But the consumption appeal, as Turner (1981) has pointed out, is limiting. It constrains the SNP in its search for votes. It makes it a *regional* national party (Table 7.17).

The other parts of Scotland consist of places that are equally distinctive to those considered so far. The Borders provides an interesting example of massive reversal to a pre-1920s pattern of political expression. The two seats in this region (1983) – Tweeddale, Ettrick and Lauderdale, and Roxburgh and Berwickshire, are the inheritors of a persisting Conservative–Liberal split. Since 1966, other parties, especially Labour, have faded in importance. In the Borders the modern Liberal party is the inheritor of the "Liberal tradition"; there have been few desertions to the SNP (Harvie 1981:147). The consumption side of nationalism has a limited appeal in this most "English" of Scottish regions. There are historic connections across the border such as migration (Hollingsworth 1970) and more recent influences such as the local TV station broadcasting from Carlisle in England. More-over, the economic and social concerns of the populace have been such that the Liberals rather than any other party vie most successfully with the Conservatives for political dominance. The sort of features of everyday life that encourage support for the Labour party in Strathclyde, indicated by large-scale public provision of housing, high unemployment, high percen-tages of the work force in heavy industries, and high levels of deprivation,

Table 7.18 Best SNP performances in 1983 predicted from 1982 regional election results.

Constituency	SNP to leading party	
	%	Party
Linlithgow	2	Labour
Clackmannan	9	Labour
Argyll and Bute	10	Conservative
Cumbernauld and Kilsyth	11	Labour
Livingston	13	Labour
Angus East	14	Conservative
Falkirk West	13	Labour
Falkirk East	15	Labour

Source: Bochel & Denver (1983 : 10).

are not present in the Borders region (Regional Trends 1981 : 31, Knox & Cottam 1981). The Labour–Catholic image of the Labour party in Scotland also probably restricts its appeal in a predominantly non-Catholic area (Darragh 1978 : 240). But the region has a high level of employment in manufacturing, specifically textiles, that creates work conditions in which opposition to the Conservatives is a likely outcome.

The Central, Fife, and Lothian regions are less clear cut. Central and Fife are in many ways most like Strathclyde in political expression and social structure. With the exception of Northeast Fife, Fife is almost exactly the same, with the exception of fewer Catholics but more miners as compensation. But over the years the SNP has become a major force in district and regional elections in the Central region. The SNP also runs very strongly in the Clackmannan constituency. The success in local elections probably reflects the "new town" and multinational issue appeal of the SNP. But in 1983 the Liberal/SDP Alliance stole much of the SNP's thunder in Central, suggesting a prevalence of "soft" rather than "hard" nationalist support.

In the Lothian region the SNP has built up a following in the Linlithgow and Livingston constituencies, but little elsewhere. In trying to predict 1983 general election results from 1982 regional election results, Bochel and Denver (1983 : 10) find Linlithgow and Livingston to be two of eight constituencies in which a gap of less than 15 percent separates the SNP from the leading party (Table 7.18). Many of the others are in the Central region. But regional votes for the SNP have not translated into parliamentary votes or seats. Elsewhere in Lothian the SNP runs poorly. Labour–Conservative contests are the norm, with the Conservatives doing especially well in Edinburgh.

Conclusion

In terms of economic base, population history, settlement characteristics, cultural attributes, and political expression, there are a number of distinctive sets of places in Scotland. There are regions such as Tayside, Grampian, and Galloway, in which contexts for Scottish nationalism, especially that represented by the SNP, are well established. But there are also regions, especially Strathclyde, in which class-based solidarities have persisted or strengthened, and in which support for the SNP is lukewarm. In the Highlands, political behavior is volatile and extremely unpredictable. In the Borders, the Conservatives and Liberals face off today as they did 100 years ago. The Central region is somewhat intermediate between Strathclyde and Tayside. Fife and Lothian are most like Strathclyde with the exception of Edinburgh and Northeast Fife. They could well be in the south of England!

Regional analysis therefore suggests that there are numerous "Scotlands." Three are particularly distinct. One in which nationalism in the form of the SNP faces a continued and probably strengthening commitment to class-based politics as this has evolved out of local experience. Another where SNP-style nationalism is and will probably continue to be well-established. And finally a third Scotland, where political commitments are, in the aggregate, fluid and volatile, but in which the SNP can expect to survive and perhaps flourish as a force in electoral politics.

8 *Place and Scottish politics*
The historical constitution of political behavior in places

The examination of aggregate political behavior is all very well for describing the reality of place-specific variations in political expression, but it is a "broad brush" treatment: it provides only a limited overview of the interpellation of place and politics. In this chapter, four places are the subject of more intensive investigation. They are Glasgow, Western Isles, Peterhead, and Dundee. These are *not* case studies of typical or average Scottish political behavior, but preliminary explorations of the links between specific milieux and political expression. They are simply different places.

Glasgow

Besides being the third largest city in Britain, the largest city in Scotland, and the center of "industrial" Scotland, Glasgow has acquired a reputation as a city with deep-seated economic and social problems. These problems are not new, but have their roots in the 19th-century industrialization that made the city. The political evolution of Glasgow parallels and to a considerable degree reflects this unhappy urban history.

Mid-18th-century estimates suggest that fewer than 200,000 people, or only 14 percent of the total Scottish population, lived in West Central Scotland (Kyd 1952). But later in that century and in the 19th century a series of industrial developments occurred – textiles, coal, iron, shipbuilding, steel, heavy engineering – that transformed the Clyde Valley into the most populous region of Scotland. This growth was partly based upon the availability of natural resources appropriate to the new technologies of metal manufacture and steam power, but the existence of a local social structure – particularly in the form of an entrepreneurial proto-middle class and trading elite – was also vital. Individual towns were reliant on particular industries for their early growth: Paisley on textiles, Coatbridge on iron-making, Greenock on sugar refining and shipbuilding, for example. But the arrival of interlinked manufacturing processes in the second half of the 19th century and associated transport networks produced an integrated industrial region (Slaven 1975). At the center of these changes was the role played by Glasgow as not only the commercial and organizational hub of the region but also as a major location of manufacturing activities in its own right.

Table 8.1 Population change in Glasgow, 1801–1951.

	Area considered Glasgow		
Year	Definition	Area ('000 acres)	Population ('000)
1801		c. 5.1	77
1811	Area approx. equal to parliamentary	c. 5.1	103
1821	burgh created in 1832	c. 5.1	140
1831		c. 5.1	193
1841		5.1	256
1851		5.1	329
1861	Parliamentary burgh	5.1	396
1871		5.1	478
1881	Parliamentary burgh	5.1	489
	Municipal burgh of 1872	6.1	511
1891	Municipal burgh	6.1	566
1901		12.7	762
1911		13.0	784
1921	Municipal burgh (enlarged)	19.2	1,034
1931		29.5	1,088
1951		39.7	1,090

Source: Cunnison & Gilfillan (1958).

The incredible scale of Glasgow's growth in the 19th century is easily recognized (Table 8.1). But equally dramatic has been the abrupt cessation of growth in this century. Up until 1901, Glasgow gained population both from in-migration and from a high rate of natural increase. After this date migration flows were outward, reducing the population resulting from the city's natural increase (Robertson 1958).

During the 19th century the largest flow of in-migrants was from Ireland, but there were other major movements from the Highlands and other parts of central Scotland (Slaven 1975). In 1841 about 16 percent of Glasgow's population was Irish-born. When account is taken of the high rate of natural increase among these migrants, it is possible to surmise that perhaps one-third of the city's population was of Irish descent in mid-century. But only about half these in-migrants were Catholics (Handley 1964); there was a heavy movement of Irish Protestants to Glasgow as well.

Many of the migrants were more or less destitute on arrival in Glasgow, particularly in the late 1840s and 1850s, seeking unskilled work in such industries as textiles, construction, iron and steel, and mining; and seeking housing in the already overcrowded tenements of the central city. A massive process of private tenement building failed to relieve the overcrowding. Appalling conditions of overcrowded and insanitary housing and periodic outbreaks of cholera, typhus, and diphtheria characterized Glasgow and Clydeside throughout the 19th century (Handley 1964, Slaven 1975, Checkland 1976).

A counterpoint to the grim and often deadly living environment of 19th-century Glasgow was the city's industrial success. The West of Scotland in general and Glasgow in particular were in the forefront of innovations in iron-making, shipbuilding, and locomotive engineering. Leading industrialist families, such as the Bairds, Elders, and Tennants, amassed large fortunes. The Corporation of Glasgow gave a public dimension to the "progressive" side of Glasgow's life by promoting public water supply, gas and electricity, electric trains and subway, art galleries, parks and an imposing City Hall. It also pursued an "urban renewal" policy, but not to much effect. There was a sharp contrast between conditions in working-class areas and those in prosperous middle-class ones. The Corporation served its masters well.

The shift from net in-migration to net out-migration from Glasgow around 1901–10 corresponds to a faltering in the city's economy. Though only apparent in retrospect – population still grew and unemployment remained low – the decline in the city's basic industries began at this time. The overseas challenges and failure to diversify noted earlier as the sources of decline started to have an impact long before World War I. However, there were limited political effects until the century was well into its second decade.

In the 19th century, Glasgow had become a city with strong, radical, Liberal politics. Apart from the elections of 1895 and 1900, Glasgow consistently returned a majority of Liberal MPs. This was the case even in the "great sectarian" election of 1886, when the central issue was Gladstone's Irish Home Rule Bill. Five out of the seven Glasgow MPs remained loyal to Gladstone, and four were returned. By 1910 only the business constituency of Glasgow Central was in the Conservative or Liberal Unionist column. The "Irish vote" certainly cannot account for this persistence. In Glasgow at least, if not elsewhere on Clydeside, the Irish *Catholic* vote was not sufficient to win elections during this period of time (McCaffrey 1970). Rather, as J. Smith (1984:34) has argued:

In Glasgow the 'commonsense' of the town was largely, though not entirely, Liberal. Most Glasgow working men believed in Free Trade, the iniquities of the House of Lords and all other hereditary positions, loathed landlordism, believed in fairness and the rights of small nations and in democracy and the will of the people. Although each individual Glaswegian's 'spontaneous philosophy' might differ – with elements of anti-Catholicism, extreme Protestantism and nationalism – liberal and radical strands were strong.

For most of the period 1885–1910, support for the Liberals crossed class lines. Increasingly, however, the Liberals became dependent on working-class support as the middle classes, for whom such issues as Irish Home Rule and the Boer War displaced traditional economic and social positions in importance, deserted the party for the Conservatives. After 1906 this working-class–Liberal alliance was challenged in Glasgow, as it had been previously elsewhere in Clydeside, by the growth of a socialist movement. This movement took various forms, but its most important organization was the Independent Labour party. The roots of the ILP in Glasgow lay in the 1880s' struggle for land reform in the Highlands and Ireland transferred to an urban setting. There was thus a strong continuity between radical Liberalism and Glasgow socialism.

The Glasgow ILP stressed the importance of elections, both municipal and parliamentary, and efficient ward organization. They also proposed a brand of "municipal socialism" as a practical and more democratic alternative to the syndicalist and revolutionary socialisms then in vogue. In this they were carrying on the traditional Liberal commitment to self-help and self-government. In this way the "radical" socialism of the ILP could be shared by both Protestant and Catholic socialists alike (J. Smith 1984 : 35–6).

The socialist movement in Glasgow was by 1909 the strongest in Britain. In that year 30,000 marched together in the May Day Parade. Many of these people were skilled workers from the shipyards, inheritors of the Liberal tradition. Others responded to the appeal of municipal socialism. As J. Smith (1984 : 38) puts it:

> Socialism became an available vision for Glasgow men and women before the First World War. Alongside the "philosophers" and "good sense" thinkers of Glasgow socialism, many working men and women came to incorporate elements of socialism into their own "spontaneous philosophy." The strong Liberal tradition in Glasgow kept in check the rise of Protestant/Catholic sectarianism and emphasized many fundamental democratic and radical principles, contributing greatly to the process through which a Liberal working class became a Labour working class. Given the ease of transition in Glasgow in the 1900s it was not surprising that all the socialists in Glasgow were, in one way or another, evolutionary socialists.

The industrial and neighborhood structure of Glasgow played a key role in the city's politics before 1914–18. From the 1870s Glasgow (and Clydeside) was the shipbuilding center of the British Empire. The city grew around the great shipyards and engineering works of Partick and Govan. In 1911 49 percent of Govan's occupied males and 42.5 percent of Partick's were employed in shipbuilding and engineering (Smith 1984 : 48). Most of these men were Protestants; there was no competition from Catholics in the skilled trades they engaged in. Nor was there much unemployment at this time to stimulate enmity and fear. Various "private associations" such as friendly societies, the cooperative movement, temperance societies and craft trade unions provided the skilled workers with locales that reproduced a "Liberal" and proto-socialist vision of the world (Smith 1984 : 47).

Elsewhere in Glasgow, the city's Catholic and highlander populations were widely scattered (Pelling 1967:403). Rather than exclusive neighborhoods or "ghettos" such as existed in other cities of Clydeside and the west of Scotland, there were limited possibilities for sectarian conflicts based on geographical segregation. In the more exclusively Protestant working-class districts, such as Govan and Partick, the possibilities of sectarianism were blunted by the heightened sense of class interest and radical Liberal philosophy predominant among their skilled male populations.

During World War I, Glasgow acquired a notoriety for political radicalism among those most exercised by the necessity for a "war effort." This image initially rested upon the Glasgow Rent Strike organized by the ILP Housing Group and ILP women in 1915 and the strike over housing evictions led by workers in the Govan shipyards where the ILP was strong. To this was added in 1919 the Forty Hour Strike, followed by a riot in the city center. The notion, or legend, as some would have it (McLean 1983), of "Red Clydeside" was born.

The war extended the predominance of heavy industries even as the techniques of mass production it generated undercut the prerogatives of skilled workers. By the early 1920s a strong adversarial relationship was established between management and workers. This coincided with a dramatic decline in world demand for Glasgow's products. It is difficult to underestimate the impact of this drastic change in economic fortunes. A first reaction of Glasgow businessmen was to "rationalize" their industries to improve efficiency in preparation for an upturn in demand. A second reaction was to diversify the economy by attracting newer light engineering industries from outside the region. Neither strategy worked. During the 1920s heavy emigration mitigated the disastrous economic situation, but by the 1930s opportunities elsewhere had dried up and unemployment reached staggering proportions. Between 1931 and 1933 the unemployment rate in Glasgow was continually over 30 percent, compared to 20 percent in Britain as a whole. As late as 1936 unemployment in Glasgow was over 20 percent (Fogarty 1945). Only World War II brought relief.

The "commonsense socialism" of Glasgow underwent a considerable transition during the interwar period. In one sense it was institutionalized. In the 1922 election the Labour party won two thirds of Glasgow's parliamentary seats. Elements of municipal socialism, particularly concerning city provision of new housing, came to characterize Glasgow even before Labour took control of the City Council in 1933. But in another sense it was transformed. Low-rent housing, city jobs and other perquisites of municipal socialism, Labourism if you will, displaced the ideological socialism of an earlier era, although of course the goals of municipal socialism had been popular all along. However, the organizations of the skilled workers had been destroyed as their industries were destroyed. In the late 1920s and early 1930s there was even a resurgence of sectarian politics as religious affiliations became weapons in the struggle for jobs. The earlier vision faded as the industrial and social circumstances in which it had been born disappeared.

Since World War II, Glasgow has witnessed a continuation of the industrial decline begun in the early years of the century. The majority of new industrial development and employment growth over the past forty years in the west of Scotland has taken place outside of Glasgow (Carter 1974). The shipbuilding and heavy engineering industries are almost dead. Movement of people out of the city into the surrounding area, particularly to the new towns of Cumbernauld and East Kilbride, has drained any of the most skilled manual workers out of the population (Henderson 1974). Glasgow today has a labor force containing one of the highest proportions among British cities of workers with low levels of skills, together with one of the lowest proportions of professional and managerial people (Checkland 1976). It also includes a population that has become increasingly dependent on municipal and national government programs that compensate for the lack of employment and other opportunities. Despite greater mobility, more mixing of social classes, increased consumption, and reduction of gross income differences, the sense of class difference is probably higher now in Glasgow than ever. With the tremendous rehousing efforts and peripheral growth of the city, residential segregation by class has increased. Out-migration of skilled workers has diminished the size of the "respectable" or upwardly mobile working class. But more importantly, the ever-present threat and widespread experience of unemployment, rising to levels in the recent past on a par with the 1930s, and the practical commitment to municipal socialism, have produced a level of support for the Labour party beyond the wildest dreams of such ILP leaders as Johnston, Wheatley, and Maxton. But it is a different "socialism" from what they had in mind. Their socialism of hope has been replaced by the Labourism of council housing and public service provision.

The Western Isles

From one point of view the Western Isles is the polar opposite of Glasgow. It is one of the most rural, most peripheral and ethnically distinct parts of Scotland. But from another point of view they are similar. Both places are "socially deprived" and have had histories of "radical politics." For those engaged in facile generalization, either viewpoint might suffice. But both "sides" of the Western Isles deserve attention.

In 1971 the population of the Western Isles was 29,891, living mostly in crofting settlements around the perimeter of the islands. By and large they do not own the land they farm but rent it from a variety of landlords, mainly absentee, under "crofting tenure." Under the Crofting Act of 1886 and subsequent legislation, the crofter's right to a given property is guaranteed, a low rent is fixed by the Scottish Land Court, and there are collateral rights to common grazings and peat diggings. The Crofting Act of 1886 was wrested from the British government of the day by the agitation of the Highland Land League. To a considerable extent it answered the grievances of the crofting populations of northwest Scotland. They were interested in acquiring tenurial rights rather than ownership. But agitation over land rights continued into the 1920s. The "Land Question" etched a permanent place in Highland folk memory. The privations of the 19th century, the evictions and clearances that predated the Land League, are kept alive in the anecdotes and tales of older people (Ennew 1980).

Persistent emigration from the islands has caused both population decline and a change in age structure. In particular, the rural areas have become more and more sparsely populated and generally older in average age. Only the town of Stornoway has experienced an increase in population. Many young people move to the mainland for work. Those who remain are faced with a severe unemployment situation. The rate of unemployment is hard to calculate because of the self-employed status of crofting and the weaving industry, one of the few alternatives to crofting. But the official figure for Lewis and Harris, the northernmost island, averaged 22.15 percent in the years 1964–75 (Ennew 1980: 48). The generally dismal picture of economic activity in the Western Isles has not changed much with the arrival of oil-related industry in the 1970s. The oil-platform construction yards failed to generate much business. Employment peaked in 1976, and by 1978 much of the workforce was redundant. This continues the history of "boom and slump" in the introduction of "new" economic activities in the Western Isles. First came the kelp, or seaweed for fertilizer, industry in the early 1800s; then, in this century, attempts at expanding commercial fishing and the commercial weaving of "tweed." In the 1970s the oil industry presented another experience of the illusion of "development" initiated from outside. Only crofting remains as "the base-line of all other economic activities. As an idea it provides the secure focus for an insecure existence" (Ennew 1980: 59).

The general level of "social well-being" in the Western Isles is much the lowest of anywhere in Scotland (see Ch. 7, Fig. 7.2, Table 7.6). Though the islands differ in the severity of social deprivation, the southern islands of Barra and the Uists have more serious housing problems than the northern islands, they share long-established consumer cost disadvantages and inferior public services (Knox & Cottam 1981). There are also notable problems of alcoholism and depressive illness traceable, at least in part, to the history of deprivation and a collective image of the past as more coherent and less stressful than the present (Ennew 1980: 110).

The political success of the Scottish National party in the Western Isles since 1970 has taken place against this background of relative economic and social deprivation. The policies of the 1964 Labour government, which promised so much to improve the lot of people in "problem regions," did not appreciably change the situation. After returning a Labour MP for the previous thirty-five years, usually with solid majorities, in 1970 the voters rejected Labour in favor of the SNP candidate. His 6,568 votes (43.1 percent) contrast markedly with those of the previous SNP candidate in the 1951 election, 820 votes (5.0 percent). So disillusionment with the Labour party, and the sitting member (the increasingly erratic and absentee Malcolm Macmillan), apparently led to the election of the ex-provost of Stornoway, Donald Stewart, as the first successful SNP candidate in a general election. Both the Labour party and the SNP can claim to be heirs to the tradition of the Land League. But by 1970, and in the context of that period, the SNP and its candidate had become more vital expressions of that tradition.

But there is probably a deeper level of attachment to the "new" MP and his party than mere disillusionment with the Labour party. SNP support comes from both ex-Labour and ex-Conservative, and probably also, new voters (see Table 7.10). There has not been a simple shift of votes from Labour to the SNP. The SNP and its candidate have tapped a "sense of place" that goes beyond the well-established picture of economic and social deprivation. In the Western Isles, the SNP uses nationalist rhetoric to channel the islanders' sense of Celtic or Hebridean individuality into resentment at those held responsible for their current condition: the English. This is something the Labour party could never do. Rather than tapping a classic class consciousness, the Labour party exploited a peasant resentment of landlords and a collective sense of deprivation. But this is better represented now by the SNP.

The Gaelic–Calvinist heritage of the islands, particularly the most populous northern ones, always had an uneasy coexistence with the tenets of socialism, however mildly construed. The other-worldly orientation of the major Presbyterian sects, the Free and Free Presbyterian Churches, discourages political activism. Yet, at the same time, the origins of these churches and their historical association with the Land League have stimulated acceptance of a "radical" if not a socialist political ethos (Hunter 1974, 1976). The appeal of the SNP is that it is a radical but not a socialist party. It can appeal to those who have voted Conservative by stressing the weakness of Scottish capital in its homeland, or the secularism of English Conservatism. It can appeal to those still exercised by the Land Question by emphasizing social justice and fairness. Finally, it can appeal to those worried about the political, cultural, and economic impacts of EEC membership by playing up its role as the only party in Scotland opposed to Scottish membership under present terms. The act of voting SNP in the Western Isles today, therefore, involves a number of interests and sentiments that were not well represented by voting Labour yesterday.

The success of the SNP in the Western Isles since 1970 is quite distinct from its success elsewhere. The life of the islands is primary. The issues and sentiments that are important to the population are distinctive. This has been reinforced by the creation of a separate regional government since 1974 (Hache 1982). The current MP must concern himself with the intricacies of crofting legislation, and make public declarations on behalf of local causes such as Sabbath (Sunday) Observance. He is the agent of the islands in the outside world. That he represents the SNP is not incidental to his success. But it is his ability to *translate* the SNP's messages into terms and issues that are locally meaningful and appear to act on them that ensures his high level of popular support.

Dundee

Dundee is an isolated industrial city on the north bank of the River Tay. Its major growth took place in the 19th century. Until the Second World War its industrial structure could be described by the phrase "jam, jute and journalism" (Pacione 1972:53). The most important was jute. Almost the whole of the British jute industry was concentrated in Dundee. The other established industries were a major publishing concern (D. C. Thomson and Co., the publishers of local newspapers, children's comic books, and such weekly or monthly papers and magazines as *The Scots Magazine*, *The Sunday Post*, and *The People's Friend*), and the manufacture of marmalade and confectionery. In fact the jute industry overwhelmed the others in the size of employment and impact on the city.

The production of jute goods began in Dundee in the 1830s after the failure of the local flax crop made the local linen industry look for an alternative raw material and new products. From this time until well after World War II the jute industry employed a majority of the city's labor force. As late as 1955, 1.23 percent of the city's productively employed worked in the jute mills (Pacione 1972:56).

Historically, a large proportion of the work force was female. In the period 1885–1923 when the industry was at its zenith, most of the jute spinners and weavers were women (Walker 1980:21). Men were employed primarily as laborers, floor-sweepers, lift-attendants, and as "hands" responsible for the preparation of raw jute. It was a "women's town," with nearly a quarter of all *married* women in paid occupations. Prior to marriage they would have been part of the "disreputable" force of mill-girls who constituted a significant subculture in the city. It was also a city that acquired a significant influx of Irish Catholics but, as Walker points out (p. 21), few Irish Protestants. This goes some way towards explaining the lack of Irish sectarianism in Dundee. But other influences were also at work. Many of the Irish in Dundee were women. This was the opposite of the case in such cities as Glasgow, Greenock, or Edinburgh. Irish women were less threatening than a large Irish male population would be. Dundee was also something of a "frontier" town as late as 1870, in which the Irish would be contemporaries of any number of other recent arrivals from the Highlands or the county of Angus rather than newcomers pushing into an already settled and stable social milieu.

Above all, however, Dundee was a Liberal city. The city fathers and the major newspaper in the 19th century, the *Advertiser*, would countenance no anti-Catholic or anti-Irish prejudices that were solely religious or ethnic (Walker 1980 : 122). In return, as they acquired the vote, the Irish men (and later the Irish women) of Dundee became a block of support for Liberal candidates. But the living and working conditions in the city were so awful and the contrasts between wealth and poverty so great that after World War I Dundee shifted dramatically away from the Liberals. This movement was partly a consequence of increased levels of membership in effective and Labour-led trade unions, especially the Jute and Flax Workers' Union. But it also reflected the efforts of the unique Labour–Prohibitionist movement associated with Edwin Scrymgeour. This movement attracted much support from the mill-girls and other women voters (Walker 1980 : 22), and culminated in the ousting of Winston Churchill in the general election of 1922.

However, the first triumph of Labour was short lived. Three months after the election, the Jute and Flax Workers' Union became engaged in a strike to protest a rationalization of spinning operations in one mill. The employers responded with a general lock-out. The locked-out workers lost sympathy with their comrades and the strike failed. From 1923 onwards, employers were more or less free to follow their own inclinations without much worker resistance and union interference (Walker 1980 : 529). Depression, competition from India, and substitute materials reduced the size of the industry and the size and nature of the labor force. As new machinery was introduced and mills introduced night-shift work, more men and fewer women and juveniles were employed. By 1931, 50 percent of the city's textile workers were unemployed (Walker 1980 : 530). The political incumbents suffered the consequence of this and the national Labour party's division into factions. In that year the Labour and "Prohibition and Labour" representation of the city was lost to a Liberal and a Conservative. As Walker (1980 : 531) puts it, "the economic gloom of the 1920s was to some extent redeemed by periodic political success. There was no such consolation in the following decade."

Since World War II the industrial structure has become much less dominated by the jute industry. By the 1960s other industries had become relatively more important. The largest of these – electrical engineering, timepieces, photographic equipment, and computers – have been located on industrial estates established by the British government. These industrial estates are situated to the northwest and northeast of the city center in the vicinity of the large council housing estates built since 1945. From 1945 until 1970, the Labour party dominated both local and parliamentary politics in

Table 8.2 Number and percentage of the working population of Dundee and Scotland unemployed, 1969–81 (July figures).

	Dundee		Scotland	
	Number	%	Number	%
1969	2,569	2.8	78,966	3.6
1970	4,406	4.9	90,598	4.2
1971	6,454	7.4	128,730	6.0
1972	6,932	7.4	136,509	6.5
1973	5,013	5.3	95,207	4.4
1974	3,783	4.0	89,840	4.2
1975	6,137	6.5	129,836	6.0
1976	7,687	7.9	165,649	7.5
1977	9,219	9.4	194,271	8.6
1978	8,890	9.1	191,906	8.5
1979	9,079	9.3	187,431	8.3
1980	11,677	11.9	236,326	10.5
1981	15,168	15.5	318,215	14.1

Source: Manpower Services Commission, Edinburgh (from Hood & Young 1982 : 112).

the city. Rather as in Glasgow, rehousing and other products of municipal socialism and the welfare state produced a large constituency for the Labour party. Also, in Dundee there was much less of the sectarian rancor that periodically emerged in certain places in the west of Scotland to dilute Labour support and feed working-class Conservatism. Catholics were fewer in number (17.6 percent in 1977), more or less assimilated; "Irish issues" were not important, and there was little or no "Orange" tradition in the non-Catholic population.

Until the 1970s, and also unlike the west of Scotland, unemployment in Dundee was very low (Table 8.2), as the new industries had compensated for the decline in textile employment. But in the 1970s there was a sudden drop in employment in what was left of the jute industry (Table 8.3). Moreover, there was a dramatic loss of jobs in the factories of Timex and National Cash Register (NCR) and at the plants of other American firms that had located in Dundee since 1945. At its peak Timex employed over 6,000 people in five factories in Dundee. By 1981 this number had been reduced to 3,900. The

Table 8.3 Textile employment in Dundee, 1911–76, as percent of total employment.

1911	1931	1951	1961	1976
48	41	23	18	8

Source: Scottish Economic Planning Department, *Redundancies in Dundee*, Research Paper No. 1, July 1979 (from Hood & Young 1982 : 111).

loss of jobs from NCR was even more dramatic: in the period 1969–71 there were 1,200 redundancies, and between 1970 and 1980 employment at NCR went from over 6,000 to around 1,000. Seven of NCR's nine Dundee factories were closed. As one local union leader commented: "we were a boom town. Now suddenly we are a doom town. It is the speed with which this has all happened which has shocked us" (quoted in Hood & Young 1982:107).

The traumatic increase in unemployment in Dundee in the 1970s, largely associated with the operations of American multinational companies, provides a backdrop for political change in Dundee during the same years. The most vivid demonstration of this is the unique success of the SNP and its candidate Gordon Wilson in capturing and *holding* the Dundee East constituency since February 1974. What is most difficult to explain, however, is why the support for the SNP is largely restricted to this one constituency rather than city-wide. The SNP has not performed well in Dundee West. Each of the constituencies has approximately 65,000 electors and shows much the same socioeconomic pattern – areas of owner-occupied housing scattered widely with large council housing estates on their northern fringes. But in Dundee East there is an area of middle-class owner-occupiers with a large amount of recently built private housing sold largely to younger buyers. Local commentators believe that it is the people in this area plus Conservative "tactical voters" and ex-Labour voters on the council estates who have provided the Nationalists with the coalition necessary to win and hold the seat. The West constituency has no key group of potential SNP voters to provide a nucleus for the party there. It is not so much unemployed workers from closed jute mills and American factories who provide *en bloc* the SNP "cadres" in Dundee. They are not confined to the east. In Dundee East, to use an American expression, the foundation of support for the SNP rests on its endangered "yuppy" population (young urban professionals), reacting to recent events such as the factory closures by turning to the SNP and away from the British parties. However, it should be stressed that the loss of jobs in the 1970s affected field engineers, computer engineers and salesmen as much as it did factory workers (Hood & Young 1982:109). In Dundee East, at least, the SNP appeal to a "new" constituency has not fallen on deaf ears. The "respectable" working class and young middle class of Dundee concerned about their and their children's *career* prospects have been sufficiently jolted by recent experiences to turn towards the one party that addresses their concerns. In this setting the SNP has an "investment" as much as a "consumption" appeal.

Until recently, many commentators have tried to explain away the SNP's success in Dundee East and the disruption of the Labour hegemony in the city as simply the result of tactical voting, a temporary reaction against the "corruption" of the Labour-dominated city government, a protest vote against Labour rather like the SNP vote in Strathclyde in the 1974 elections, or solely the vigor and public visibility of SNP's MP. They are now at a loss. The 1983 election in Dundee East, at a time when there was a marked national downturn in SNP performance, indicated a persisting attachment to the SNP and its candidate. Perhaps, this is the latest manifestation of the collective idiosyncrasy that previously produced the Scottish Prohibition party and its unrealistic drive to an alcohol-free promised land. More likely, it is a collective response of those not set in their political ways to recent economic traumas in their own lives and that of the city. This is not a protest vote. It is a cry for change.

Dundee has always been a city apart from the rest of Scotland, neither of the northeast nor the industrial center. This detachment comes out in both its economic and its political history. The tradition continues.

Peterhead

The town of Peterhead is located about 42 km north of the city of Aberdeen. But it is relatively inaccessible by road, and until the 1970s was isolated socially and culturally. Only its prison, notorious for its population of "hard-timers," was well known in the rest of Scotland. The town itself, as Moore (1982:2) describes it, "was probably only known to seafarers and to connoisseurs of Protestant sectarianism."

The economic life of the town has always depended on its harbor. Even the prison was built to house convict labor engaged in harbor work. In the early 19th century Peterhead was a major whaling port; by the 1870s herring fishing had become more important. Today, it is still a center for North Sea inshore fishing, but with reduced numbers of fishermen and local boats (Moore 1982:61). Periods of prosperity in the fishing industry, such as the 1970s, have been matched by periods of destitution, such as the 1930s and 1960s. Though Peterhead, now a town of 14,500 people, remains dependent on fishing and fish processing, other economic activities have become increasingly important. There are a number of branch-plant factories operated by such firms as General Motors, Clarksons Tools (Thorn), and Crosse and Blackwell (Nestle), and manufacturing is now a more important source of employment than fishing (Table 8.4).

Peterhead is one of the nearest points on the British mainland to the North Sea oil and gas fields. Given this proximity it is not surprising that after the major oil discoveries of the early 1970s, Peterhead and its harbor would become transhipment points for the new industry. Offshore oil operations require substantial support facilities. To sustain the necessary supplies, a base with deep-water jetties, warehousing, and cranes is required. But once in place, such a base is not a large employer. Moore (1982) has shown that the direct employment effects of North Sea oil in Peterhead have been limited. Many jobs proved to be temporary. Some firms brought in workers from

Table 8.4 Peterhead: employed males and females, by industrial Sector.

	1971		1972		1973		1974		1975		1976		1971-6	
	M	F	M	F	M	F	M	F	M	F	M	F	M	F
agriculture and fishing	715	79	724	68	656	78	616	70	535	63	531	65	−184	+ 14
manufacturing	1,910	1,047	1,854	1,341	1,988	1,455	1,891	1,392	1,872	1,069	1,877	1,137	− 33	+ 90
construction*	508	31	547	34	547	40	547	53	547	81	541	141	+ 33	+110
services	1,672	1,605	1,645	1,675	1,707	1,776	1,691	2,004	1,865	2,078	1,847	2,255	+175	+650
total	4,805	2,762	4,770	3,118	4,898	3,349	4,745	3,519	4,819	3,291	4,796	3,598	− 9	+864

* Corrected to allow for migrant construction workers.
Source: Moore (1982 : 116).

elsewhere. The major impact has been on the expansion of banking services and pubs and restaurants. In the short run, the prison, the established factories, and the fishing industry were affected by high labor turnover. In the long run, however, it is these industries which, in default of petrochemicals, will provide the basis for long-term employment.

Peterhead gives the appearance of being a fishing community despite its mixed economic base. The local population cultivates this image of the place and themselves (Bealey & Sewel 1981 : 149). "Peterheadians" pride themselves on "the qualities of independence and individualism derived from fishing and characteristic of the whole population" (Moore 1982 : 78). That there is considerable truth to this self-image is indicated by the resistance of many to the acceptance of social welfare benefits, opposition to trade unions, and penchant for founding and splitting fundamentalist Protestant sects. But Peterhead is decidedly not a single or simple folk community. The ties with a wider world of international business and finance have already been mentioned. Within the town there are numerous voluntary associations and churches neatly placed in a hierarchy of residential districts. One of the clearest distinctions is between fishing people and the rest.

Until the reorganization of local government in 1974, Peterhead had a Burgh Council of 12 councillors from whom were appointed magistrates and representatives to the Aberdeen County Council. There were no parties in local politics. Councillors adhered to the idea that Peterhead was "a unitary community in which the people's representatives could agree by a process of rational discussion – even when there was no rational basis for agreement" (Moore 1982 : 81). Bealey and Sewel (1981) report that the lack of parties led to confusion and incoherence in the conduct of council business. "Non-decision making" seems to have been the council's forte. Councillors consistently adhered to the view that there were no differences of "interests" between them or among their constituents to be served by party affiliation. In the absence of organized politics in local government, conventional policies were rarely challenged. They generally favored the town's small proprietors and nonmanual workers. For example, Peterhead's council rents are among the highest in Scotland and its rates (property taxes) among the lowest.

There was not much base for a Labour party in Peterhead. The fishermen were and are fiercely antiunion. The engineering workers had trouble enough amongst themselves to prevent extending union efforts into town politics. There was resistance to unionization from many workers, particularly those in the Protestant "Brethren." Adopting partisan and combative positions in public was not popular. Council tenants, though 45 percent of the potential electorate, were not a good base. Councillors were elected at large, so a "council house" or Labour candidate running on housing issues could always be represented as a threat to rate-paying owner-occupiers (Moore 1982 : 84). Yet there was a widespread and deep distrust of the council (Bealey & Sewel 1981 : 186). This also worked the other way. As Moore (1982 : 84) reports on his field work in Peterhead: "In talking to ex-Councillors I was very struck by the depth of hostility to council tenants, rooted in most cases in the belief that people should be independent and own their houses, but in a few cases rooted in what one can only describe as outright hatred."

However, at the same time that working-class interests have not been organized, "antiworking-class" interests have not been well organized either. The success of local small business and property interests rested largely on the inaction of others rather than their superior organization. Particularly in parliamentary politics, the local Conservative party, covering the East Aberdeenshire constituency until 1983 and Banff and Buchan since then, has been ill organized and inefficient, if largely successful (Moore 1982 : 85). In Peterhead the political base of the burgh council rested on social networks established by individual councillors. Voting was on a personal basis. Different councillors appealed to the votes of particular churches or ex-servicemen (the "British Legion vote"), according to their own affiliations and acquaintances. Six voluntary associations provided the major links that bound together voters and support groups: Rotary, Round Table, Professional and Businessmen's Club, and three churches (two Church of Scotland and one Episcopalian). These groups in turn were strongly tied to local firms and small businesses (Bealey & Sewel 1981, Moore 1982 : 85–8).

It is in the context of a "depoliticized," if in practice decidedly partisan, local government and the peculiar local culture of "sturdy independence" that the tremendous political changes of the 1970s must be set. Since the replacement of the burgh council with a district council in 1975 and direct election to the new regional councils, party labels have invaded Peterhead. Councillors are no longer always "Independents." The most important label, following on the heels of success in the 1974 general elections, is that of the SNP. The SNP won East Aberdeenshire in 1974 with a 13 percent majority. A considerable amount of support for the SNP came from, and comes from, Peterhead. Indeed Dyer (1981 : 62) has dubbed Peterhead "the centre of Grampian nationalism." In most recent district and regional council elections, Peterhead has returned only SNP members (Table 8.5). These local politicians and the SNP's local MP (1974–9) and candidate (1979–83), Douglas Henderson, are of a decidedly conservative bent. They are in the "Tartan Tory" wing of the party. They have inherited the mantle of the old burgh council.

The SNP appeals in Peterhead, both locally and in parliamentary elections, *across* class lines. It consequently taps the "nonpartisan," "nonideological," antiunion and free enterprise sentiments so powerfully present in the town. At the same time it appeals to the interests of the "community" as a whole against the changes and unwelcome trends coming from "outside."

Workers who came to the town as part of the "oil boom" have not affected

Table 8.5   Election results for Peterhead council seats, 1980 and 1982.

A. BANFF AND BUCHAN DISTRICT COUNCIL ELECTION 1980

Meethill/Boddam		Turnout 29.9%	
S. Coull	SNP	1,161	79.6%
A. Taylor	Con.	294	20.2%
Clerkhill		Turnout 32.4%	
D. J. Mackinnon	SNP	531	54.0%
T. Nangle	Con.	230	23.4%
D. Carnie	Ind.	219	22.3%
Kirkton/Roanheads		Turnout 32.7%	
L. Johnson	SNP	377	48.6%
G. Baird	Ind.	303	39.1%
B. Gall	Ind.	91	11.7%
Buchanhaven/Catto		Turnout 35.7%	
J. Ingram	SNP	859	76.5%
S. M. Drumsfield	Con.	263	23.4%

B. GRAMPIAN REGIONAL COUNCIL ELECTION 1982

Peterhead East		Turnout 33.4%	
J. Ingram	SNP	951	45.8%
R. Warrender	Con.	466	22.4%
M. D. Buchan	Ind.	434	20.9%
P. Seed	Lab.	209	10.1%
Peterhead West		Turnout 34.5%	
S. Coull	SNP	1,159	49.8%
R. T. Antczak	Con.	709	30.5%
J. D. Towers	Lab.	419	18.0%
B. Gall	Ratepayer	30	1.3%

Source: Bochel & Denver (1980, 1982).

this political outlook. Though many are union members they have remained distant from local coworkers and local politics. Partly this reflects continued affiliation with places of origin. But it also reflects skill division, with incomers often more highly skilled, and the constant threat of deskilling in conditions of rapid technological change. One worker's threat is another worker's opportunity. In such circumstances, labor solidarity is the exception rather than the rule (Moore 1982:182–3).

In the long run, however, the SNP in Peterhead is faced with a major dilemma. Its local appeal is that it stands for the status quo rather than change. Yet nationally the SNP stands for Scottish independence, with the specter, if you are standing in Peterhead, of a Scotland dominated by "the men from Glasgow" (Naughtie 1976). The low turnouts, even by local standards of generally low turnouts in elections of all varieties, and lower than expected "yes" vote in the referendum on a Scottish Assembly (June 1979) surely suggest such a dilemma (Bochel *et al.* 1981).

The northeast of Scotland, especially places such as Peterhead and other fishing or small towns, provided the heartland of SNP success in 1974. For 40 years previously, parliamentary politics had been dominated by the Conservatives, the controversial Bob Boothby was MP for 34 years, and local politics was "nonpartisan." The SNP now offers an alternative. It is the "first" party in Peterhead, the "second" in Banff and Buchan as a whole. But the roots of its success lie with what Turner (1981 : 90) calls "the romantic myths of rural and small town Scotland" or what I have termed the "cultural–consumption" appeal of nationalism. It is yet to be seen whether this can be combined with an appeal to the material interests of those hitherto excluded from Peterhead's political life. If election turnouts are any measure of the political participation of the excluded, they are still staying home on election day. So far *they* have neither candidates nor party.

Conclusion

The interpellation of place and political behavior has been traced for four different places. These vignettes *illustrate* the place-specific appeals of different political parties, the links between the changing fortunes of places and parties in places, the different meanings attached to different parties in different places, and the dynamic nature of place–political identity interaction. They thus *support* the view that the intricacies of political behavior cannot be divorced from consideration of place. It is clear, for example, that the appeal of the SNP in Dundee is quite different from its appeal in the Western Isles or in Peterhead or in Glasgow. It is also apparent that socioeconomic change can produce different political responses in different places. The existing "sense of reality" in places constrains and channels the possibilities for responding politically to widespread or local socioeconomic change.

9 *Place and Scottish politics*
Place and political mobilization

Scotland is a particularly appropriate unit for considering the links between place and political mobilization in the form of integration into a larger state (Britain) and the interplay between class and ethnic identity.

By many accepted views of the term, Scotland constitutes a "nation." It is culturally distinctive, was until 1707 a politically independent state, occupies a distinct piece of territory, and has access to the outside world other than through the core territory of the state of which it is presently a part. What is surprising is that, until recently, distinctiveness and credentials for viability did not engender political nationalism. For most of its modern history, Scotland has been able to exist with benign multiple identities. Class-consciousness, Britishness, and Scottishness have coexisted. But they have coexisted in different ways in different places. Political mobilization around Scottish identity has become possible only because of the confluence between changes in the socioeconomic fortunes of places, on the one hand, and the expanded activities of an organization, the SNP, on the other, that appeals for support by accounting for those changes in nationalist rather than conventional left–right terms.

Place and national mobilization

Since the 19th century, Scottish nationalism as an organized movement rather than simply a political sentiment has come in three phases (Urwin 1977). In the late 1800s the traditional channels of Scottish political influence in London, largely the prerogative of the landed gentry, went into decline. In combination with the expanded role of central government in Scottish affairs, this led to complaints about the neglect and mistreatment of Scotland. Various types of administrative devolution, particularly the creation of the Scottish Office, were a response to the criticism. Cultural nationalism was limited in popular support, even though it was prominent in most other European nationalisms. Attempts to make a distinctive Scottish culture or "Scottishness" the basis for a political movement came only after World War I. But even then it did not strike a responsive chord in urban-industrial Scotland or fit with the "patron–client" or "independent peasant" politics of rural Scotland. Its proponents, who in the early 1930s formed the SNP from a number of smaller movements, were widely regarded as "lunatic fringe" or "crank" politicians. All efforts to construct a cultural ideology that would attract large numbers of voters failed. Partly this reflected the limited appeal of the Gaelic enthusiasms of many nationalists. But it also reflected the lack of a coherent and widely accepted definition of Scottishness.

The third phase, dating from the 1960s, is one in which ethnic identity, rather than expressing a set of cultural interests, has been defined as setting the social and territorial limits to a set of political–economic interests. The SNP has shifted its focus away from cultural nationalism towards the ethnic–territorial (anti- or non-Scottish) bias of contemporary economic and political management. In this incarnation the SNP has managed to expand its appeal and mobilize activists and supporters on its behalf. It has done this more by "remobilizing" the already active population of voters and activists than by mobilizing hitherto untapped groups of supporters. The potential for doing this, however, has varied from place to place.

The "modern" SNP strategy has been to avoid competition in terms of "left" and "right" by portraying Scotland as a *people* and *nation* systematically exploited and denigrated by London governments and English businessmen. Activists who have tried this have been expelled from the SNP publicly and vociferously. SNP leaders and activists, then, have tried to change the agenda of and modify ways of thinking about Scottish politics. They have been engaged in countering the conventional "mobilization of bias" engendered by the two major parties, both of which have had, with only one or two exceptional periods and individuals, a consistent hostility towards viewing Scottish problems in "Scottish" terms. They have preferred to frame their appeals and policies in terms of class (the Labour party) or the British nation (the Conservative party, long officially known in Scotland as the Unionist party). Mobilization was previously on their terms or not at all.

The Scottish situation with respect to political mobilization accords closely to the argument of Rokkan (1970: Ch. 3) and Hirschman (1979). They argue that mobilization is stimulated by crises (Hirschman) or sequences of events (Rokkan) that have different impacts in different kinds of social environment. Changes in political direction and content are favorably received in certain settings and give rise to new forms of political expression or "voice" as social change mobilizes the previously unmobilized or "remobilizes" the previously mobilized in new ways as they "exit" from their previous political attachments.

In the early part of this century, political mobilization in Scotland took the form of identification with and attachment to the Conservative and Labour parties. As affiliations shifted from the previously hegemonic Liberal party, the Conservatives picked up considerable support in rural Scotland, and the Labour party almost completely displaced the Liberals in the heavily populated West Central region. The appeal of the Conservative party was to

Table 9.1   The performance of the major and "third" parties in general elections in Scotland, 1955–83.

	Con. %	Lab. %	Con. + Lab. %	Lib. %	SNP %	Other %
1955	50.1	46.7	96.8	1.9	0.5	0.8
1959	47.2	46.7	93.9	4.1	0.8	1.2
1964	40.6	48.7	89.3	7.6	2.4	0.7
1966	37.7	49.9	87.6	6.8	5.0	0.6
1970	38.0	44.5	82.5	5.5	11.4	0.6
1974 (Feb.)	32.9	36.7	69.6	7.9	21.9	0.6
1974 (Oct.)	24.7	36.2	60.9	8.3	30.4	1.2
1979	31.3	41.9	73.2	8.5	17.2	0.8
1983	28.4	35.1	63.5	24.5	11.8	0.3

Sources: Author's calculations, based on Craig (1971) and *The Times* (12 October 1974), *Daily Telegraph* (5 May 1979) and *The Times* (10 June 1983).

stability, maintaining the established order in land ownership and religion, and preserving the imperial status quo. In many areas where the Conservatives became dominant, however, there was a persistent oppositional vote. But this never crystallized around the Labour party. The Labour party had too many negative connotations, its associations with Glasgow, trade unionism, Catholicism, and secularism, in whatever combination, to achieve widespread support as a successor to the Liberals.

On Clydeside, and in the West Central region in general, there was a dramatic shift to Labour after World War I. But this mobilization was attenuated by religious divisions that restricted the building of class capacity and the formation of class-consciousness. Only since World War II have these divisions receded, largely under the practical influence of municipal socialism and the rise of the welfare state. The Labour party has been the main beneficiary of class mobilization in Scotland. Other parties, such as the ILP, maintained an independent existence for many years, but though initially more radical, came increasingly to share the program and policies of the Labour party.

For the ten years after World War II, third parties did badly in Scottish politics. By 1955, 97 percent of Scottish votes went either to Labour or to the Conservatives. But since then both Labour and Conservative shares of the vote have steadily eroded away (Table 9.1). The 1974 elections saw the rate of change quicken. Over the past thirty years, the Conservative vote has halved, from an absolute majority in 1955 to just over a quarter in 1983. The decline in the Labour vote was concentrated in the early 1970s and in the early 1980s. But a downward trend is clearly evident – as is its cause.

The recent electoral successes of the SNP and the SDP (Social Democratic party)–Liberal Alliance have been remarkable. It is too soon to tell whether the 1983 success of the SDP–Liberal Alliance marks the resurgence of a new Liberalism in Scottish politics or a temporary home for disaffected supporters of other parties. Until 1974 there was a similar ambiguity in interpreting the early successes of the SNP. But in October 1974 the SNP increased its vote to 30.4 percent of the total Scottish vote, replaced the Conservative party as the second largest party in terms of votes, won 11 seats, came second in a further 42, and appeared to have broken the back of the two-party system. This level of success in Scotland as a whole did not of course continue. But at least in some parts of Scotland, the SNP was established as a major party. A major change in partisanship had occurred. What had perhaps started as a protest movement in the late 1960s had become linked to an assertion of ethnic identity. Even in Glasgow this seemed to be the case (Brand & McCrone 1975). During the period of SNP decline between 1974 and 1979, a preference for SNP policies, particularly economic ones, persisted among its core supporters. Those who abandoned the party in 1979 were those who, discontented with the performance of the major parties in 1974, had given the SNP "a try" (Brand et al. 1983). But in 1979 they were still searching. Perhaps they are the source of many SDP–Liberal Alliance votes in 1983.

The elections of the early 1970s clearly indicated the mobilization of a nationalist vote in Scotland. Even attempts to explain it away have succeeded in pinpointing a Scottish, if not an independence, identity amongst SNP supporters (Brand et al. 1983). But each account fails to distinguish the different sources of SNP support in different places. The economic and political transformation of the 1970s – from the perceived failure of Labour regional policies in the late 1960s, to the economic downturns of the middle 1970s, the oil boom in northeast Scotland, the growth of a corporate branch-plant economy, and the perceived impact of EEC membership – had different impacts in different places. In some places they joined with previously loose political affiliations and the attraction of SNP ideology to produce SNP voting. Elsewhere the response was more a disaffection from the major parties, indicative of an increase in issue – and a decline in habit-voting rather than a permanent partisan shift (Brand et al. 1983).

Support for the SNP increased or was maintained in the 1970s in places where electoral participation did not increase. Turnouts in its "core" areas such as Tayside and Grampian are much the same as before SNP intervention, or even down somewhat. This suggests that the SNP remobilized existing voters rather than brought in large numbers of new voters. There is something of a "generational" effect at work, however. SNP spokesmen and local politicians believe that they attract relatively more of their votes from younger voters. Survey evidence supports this claim (Hanby 1977, Miller 1981 : 147). The appeal of the SNP, then, has been at least in part, to younger people in places where that appeal has resonance.

The pattern of SNP organization and candidatures has been critical, however, in stimulating and tapping support. Although in existence since

Table 9.2 The electoral history of the SNP, 1929–83.

Election	Number of candidates	Number elected
1929	2	0
1931	5	0
1935	7	0
1945	8	0
1950	3	0
1951	2	0
1955	2	0
1959	5	0
1964	15	0
1966	23	0
1970	65	1
1974 (Feb.)	70	7
1974 (Oct.)	71	11
1979	71	2
1983	72	2

Sources: As for Table 9.1

Table 9.3 Seats won by the SNP in October 1974, and previous SNP activity.

Seat	Candidate in 1966	No. of SNP branches in 1966
Dundee, East	no	0
Aberdeenshire, East	yes	3
South Angus	no	1
Argyll	no	3
Banffshire	no	2
Dunbartonshire, East	yes	4
Western Isles	no	1
Galloway	no	0
Moray and Nairnshire	no	0
Perth and East Perthshire	yes	3
Clackmannan and East Stirlingshire	yes	6
		average for all Scottish seats: 2

Sources: Craig (1975), Brand (1978 : 289–91).

the 1920s and having contested its first election in 1929 (as the National Party of Scotland), only since 1964 has the SNP contested fifteen or more Scottish seats (Table 9.2). The electoral history of the SNP, then, is as much a history of party expansion as a history of increases in party support. There is evidence that the party never successfully targeted its constituency until the 1970s. SNP candidates appeared first in Glasgow and Renfrewshire, spreading chiefly to other seats in the central section before 1970. Only after 1970 did they begin to contest rural seats. In the October 1974 election, most of the seats won by the SNP were ones in which the party had not been particularly active before 1966 (Table 9.3). Before 1966 the party did not build up support in specific seats, a strategy favored by the Liberals. Rather they fought a range of seats, but few consistently (Taylor 1973 : 45). They often withdrew after intervening. Potential support, therefore, even in the presence of crises or a shift to issue voting, was never successfully tended or tapped. The "organizational capacity" of Scottish nationalism was radically insufficient until the 1970s.

There is evidence that until the 1970s, SNP activists were divided over the strategy they should follow in making their appeal and building up support. Before the 1920s nationalist groups had been nonelectoral in strategy, and attempted to influence the main political parties through pressure group tactics. The failure of this strategy led directly to the formation of the SNP, based on what McAllister (1981) calls "a differentiated organization and an electoral strategy." But the choice between this approach and that of "communal permeation," building from the grassroots through organizations other than a party, was never fully resolved. This lack of resolution dogged the party for many years. In 1942, for example, a split in the party led to the formation of the Scottish Convention, a pressure group which sought the support of all favoring a Scottish parliament. It explicitly rejected electoral competition. Other splits reflected the same issue even when bound up with other questions such as Home Rule versus Independence, and the place of cultural and literary enthusiasms within the party. In 1948 the party established an organizational framework that represented the victory of the electoral strategy and the ascendancy of political over cultural aims. Though an annual membership meeting was established to determine party policy, individual branches retained considerable autonomy.

Vertical communication within the SNP through center–branch links remained weak until the early 1960s. In 1962 a party report advocated a number of changes that strengthened party organization (McAllister 1981 : 244). These included a clearer definition of responsibilities for party office-holders, more emphasis on stimulating branch growth, and the creation of a small executive committee. A national organizer was also appointed at this time who carried out many of these changes in 1964 and 1965. Between this time and 1970 the SNP was, in Webb's (1977 : 103) words, "transformed into a modern, efficient mass political party."

The decentralized and "bottom-up" nature of SNP organization, however, did not change. In the period 1974–9, when the party had eleven MPs, this not only allowed for a differentiated appeal in different places, for "Tartan Tories" in some places and "Scottish Socialists" elsewhere, it also created tensions and conflicts. These conflicts emerged publicly in disputes over whether the SNP should support the Labour government's devolution proposals, whether MPs should take precedence over the National Executive Committee in parliamentary affairs, and whether party HQ should enforce party policies and candidates in the face of local recalcitrance (Kauppi 1982). Countermobilization by the other parties focused on these divisions and was paralleled by renewed efforts at organization on their part (Kauppi 1982).

From 1979 to 1982 the SNP was riven by conflicts over ideology and program. The old division over electoral strategy reemerged. A new division emerged over "the class issue"; whether it was possible for the SNP to become a truly national party unless it could make inroads into Labour support in West Central Scotland (Grasmuck 1980). The problem here is that the "small town populism" that has formed the core of the SNP appeal in its "heartland" is at odds with the appeal that would be necessary for success in Glasgow and its vicinity (McLean 1977, Turner 1981).

That a combination of nationalism and socialism is not simply a fanciful possibility, however, was demonstrated in the mid-1970s. In 1975–6 a number of prodevolutionists within the Labour party in Scotland, including two MPs, broke with the party and formed the Scottish Labour party (SLP). This party never achieved much in terms of electoral support or political impact (Drucker 1979, Mair & McAllister 1982). But it assumes some importance in its attempt to draw support in terms of both class and ethnic–territorial identity. The SLP was a socialist party, but distinctively nationalist in its commitment to substantial devolution or more. The failure of the party to survive the 1979 election suggests the limitations of the syncretic appeal. But organizational problems, in particular the identification of the party with one of its two MPs – Jim Sillars, MP for South Ayrshire – and "entryism" from the far left, undermined the party almost from the start (Drucker 1978).

What is clear is that the secret of the SNP's success in the 1970s – its ability to offer a differentiated appeal in different places – is no longer possible (Kauppi 1982). Scotland-wide debates over party organization and ideology have focused attention on the party's central aims and policies. To move in the direction of the SLP, though, is to probably abandon the solid core of "ethnic voters" brought to the party by organizational efforts of the 1960s and 1970s. That is the crux of the mobilization dilemma facing the SNP in the 1980s (and also other separatist parties, such as Plaid Cymru in Wales).

Translating social or national identity into political demands by means of electoral competition is not a mechanical and inevitable process. Though political environments are the settings for mobilization, there must be some form of organization to facilitate the process. In the 1970s the SNP proved able to stimulate and exploit in certain places sentiments of anti- or non-Scottish bias in economic and political management. Prior to this time it had been unable to build sufficient organizational capacity to exploit such feelings and interests. The problem the SNP now faces is no longer organizational. It is also rather like those faced previously by the other major parties. Production and consumption cleavages defined geographically within Scotland, along with persisting religious and other ideological divisions, limit the possibilities for a nationwide appeal. So even for a party which bases its appeal around *Scottish* themes and interests rather than British or class ones, place or locality continues to set limits to partisanship.

Conclusion

The purpose of Chapters 7 through 9 has been to illustrate and support the place perspective by reference to a concrete example, recent Scottish politics. There was no intention to offer an encyclopedic history of Scottish political behavior. From three directions – aggregate political behavior, the historical constitution of political behavior in places, and political mobilization (or, if one prefers, Scotland as a whole, specific places, and political parties) – an argument has been made for the interpenetration of place and Scottish politics. The point here is not that much of the information presented has not been previously available. It is more that it has never previously been placed in this framework. The framework implies that "mass" political behavior is intrinsically geographical, structured by the central elements of place identified in Chapter 3. But the argument rests on exposition rather than conviction. The reader should draw his or her own conclusions about the efficacy of viewing Scottish politics in terms of place. My view is that by comparison, explanations such as uneven development and population composition are inadequate to the task. Neither empirically nor conceptually are they up to the job of explaining the complexity that Scottish politics illustrates. They are also the only other serious alternatives.

10 *Place and American politics*
Aggregate political behavior

The skeptic might say that illustrating the place perspective by using a Scottish case study does not provide a sufficiently demanding "test." In particular, Scottish politics involves clear ethnic–territorial issues that lend themselves to interpretation in terms of the thesis developed in this book. A distinctive "Scottishness," or ethnically enhanced *sense of place*, combined with marked *locational* disparities within the country create a *bias* in favor of the perspective.

Whatever the truth in this argument, and I believe it is of limited validity, the United States certainly provides a more critical setting for evaluation. As noted earlier, the United States has been both the "homeland" of the nationalization thesis, and one of the most frequently invoked examples of its veracity (Schattschneider 1960, Stokes 1965). This view is expressed in the belief that the United States *has become* the most mobile, individualistic, liberal, and modern society of all. Once upon a time it was none of these. Even if places might survive in Europe, in America there is only "geometric space" containing a "mass society" (Le Lannou 1977).

This is quite a change from the image of America first given coherence by Tocqueville. To Tocqueville, America, unlike France or other states with a history of feudal, political hierarchies, created its political life out of the fabric of everyday and local political "customs" (Birnbaum 1970). The state grew out of society, so to speak (Suppa 1984). National political institutions grew out of a vital local life, a propensity to make and join local political associations – and an ideology of liberal individualism. That is to say, and to parody the modernization theorists, America was "born modernized." It did not need to become so! At least in this respect Tocqueville was probably correct. In comparative terms, American local government is still strong, and local autonomy is still of considerable importance in American folk-ideology (Suttles 1972, Peterson 1979, Clark 1985).

Unless one persists in confusing place with community, there is in fact no need to see the dominant American liberal political ideology as incompatible with the place perspective. The ideology itself can be seen as a set of ideas molded by a particular historical geography of settlement, penetration, and exploitation (Hartz 1955, Burnham 1984).

However, there is a real tension in the United States, as elsewhere, between the centralizing state and local powers. This tension has lain at the center of American political history. It has been present since "the founding." And it has led both to continued attempts to capture the apparatus of central government for local advantage and to secessionist threats.

The main argument of this chapter is that place is of both historical and continuing importance to American politics. Under the policies of the New Deal (itself the product of a historic compromise between local interests), and as a result of the growth of an American welfare state, place-to-place differences diminished somewhat. But only the fruits of a continued American domination of the world-economy could guarantee this (Agnew 1987a). Since the late 1960s, as a result of a diminution of this dominance, and other trends, these differences have returned.

The American case study is organized in three chapters, following the categories used for the Scottish case. This chapter, on aggregate political behavior, provides an overview and attempts an explanation of regional and place-specific patterns of electoral behavior from 1880 to the present. The second chapter examines the links between place and political behavior in four specific places. Finally, a third chapter explores the connection between place and political mobilization.

Aggregate political behavior

ELECTORAL BEHAVIOR 1880–1984

American electoral history lends itself to geographical analysis of one sort or another. The recent literature on American national elections, especially presidential ones, has divided the elections since the late 18th century into five party systems. Burnham (1982: 102–15) refers to these as the "experimental system, 1789–1820," the "democratizing system, 1828–1860," the "Civil War system, 1860–1893," the "industrialist system, 1894–1932," and the "New Deal system, 1932–?". In the two earliest periods, a politics of "sectional (regional) compromise" prevailed; whereas from the Civil War through the years of the New Deal, a politics of "sectional dominance" was the rule (Archer & Taylor 1981). Since the late 1940s, patterns of party support have become much less predictable and each election *seems* to be a critical one. Archer and Taylor (1981:36) refer to this period as one of "sectional volatility."

The formal organization of electoral competition in the United States has long followed "duopolistic" lines. Two large, ramshackle and adaptive political coalitions, the Democratic and Republican parties, have dominated American local, state, and national politics for over a century. It is these parties that have adapted to and exploited the shifting tides of local and regional interests and sentiments. In the politics of sectional compromise, the parties of the day *managed* conflict, but after the Civil War this same conflict was mobilized and exploited by the parties (Archer & Taylor 1981:36). Only through managing or manipulating such conflict can parties organize the electorate to achieve national (or other) control. At the same time this constrains the choices facing the electorate. As Schattschneider (1960:58) observed:

> The parties organize the electorate by reducing their alternatives to the extreme limit of simplification. This is the great act of organization. Since there are only two parties and both of them are very old, the veterans of a century of conflict, it is not difficult for people to find their places in the system ... Most Americans are veterans of the party wars; they know where they belong in the system.

Since the 1880s the two parties have seemingly reversed their regional constituencies and bases of support. At that time and through until 1928–32, the Republican party was at its strongest in the northeastern United States. The Democratic party, to the contrary, was at its strongest in the South and the West. Since 1932 these positions have been switched, although there has been much more volatility and local variation since 1948–50 than there was before (Phillips 1969). The evolution of these historical patterns is now described in more detail.

It can be plausibly argued that the Civil War and its aftermath constituted the only real revolution in the history of the United States (Moore 1966:111–55). The party of the Union and the North, the Republicans, triumphed in 1865. Until 1872 the absence of Southern access to national political institutions guaranteed a free hand to the Republicans. Not only was slavery abolished, but the Republicans set in motion policies that inaugurated a new era in American economic development. By 1874 this revolutionary phase had run its course. In that year, with the re-entry of the South into the political system, a dramatic improvement in the fortunes of the Democratic party produced a "partisan deadlock" that lasted from 1874 until the "capture" by the Republicans of all branches of the national government in 1896 (Burnham 1982:107).

The period 1874–94, however, is critical in American political history. During this time both parties came under the control of elites who favored both industrial development and private enterprise. Industrialization after the Civil War was no longer as divisive as it had been before it. One cause of this was the emergence of two major "underclasses" who threatened the new order: the farmers, especially in the cash-crop Southern and Western regions and the immigrant-based, and hence ethnically fragmented, urban proletariat. But the parties, especially and most successfully the Republican party, played these groups off against each other. So although in the 1880s there was a distinct possibility of a populist–socialist alliance against the established parties, this had dissipated by 1894. Burnham (1982:109) explains this failure as follows:

> First, the urban working class was too immature and fragmented internally to work effectively with the agrarian rebels. But more than

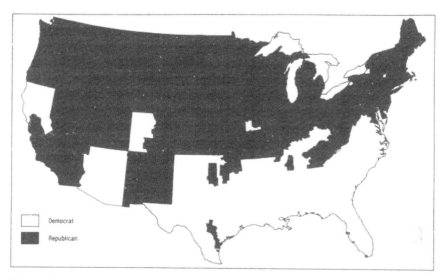

Figure 10.1   The geographical distribution of congressional seats by party, 1921–3.
Source: Bensel (1984 : 376).

this, there appears every evidence that the combination of the "Democratic depression" of 1883 and severe ethnocultural hostilities between new-immigrant workers and old-stock agrarians created an urban revulsion against the Democrats which lasted into the late 1920s.

The pattern of political alignments that emerged in the 1890s gave rise to three sectional systems: a solidly Democratic South, an almost equally solid Republican Northeast, and a "colonial" West in which neither major party was dominant but which spawned a variety of protest movements against both of them. The sectionalism was extreme. For example, excluding the complex election of 1912, 84.5 percent of the electoral vote for Democratic presidential candidates between 1896 and 1928 came from the South and the border states of Maryland, Kentucky, West Virginia, Missouri, and Oklahoma. In 1920, only about 12 million out of 105 million Americans lived in states in which they had a choice between two major parties of which either could win. In congressional elections, of the 51 congresses between 1880 and 1984, the parties were the most regionally polarized in the Sixty-seventh Congress (1921–3) which convened after Harding's landslide presidential election in 1920 (Fig. 10.1). In general outline, however, the distribution of seats between the two parties in the Sixty-seventh Congress was typical of the Republican-dominated House of Representatives of the late 19th and early 20th centuries (Bensel 1984 : 374).

The election of 1928, however, brought a huge new block of immigrant and female voters into the electorate, and as a consequence has been

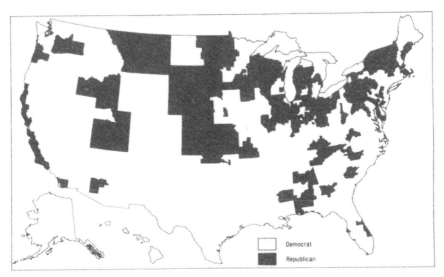

Figure 10.2 The geographical distribution of congressional seats by party, 1965–6.
Source: Bensel (1984 : 377).

identified as a critical "realigning" election in the Northeast (Andersen 1979). Many of the new voters were, in Andersen's terminology, "doubly nonimmunized" in political terms "for they had no personal political experience and had not 'inherited' a party identification from their immigrant parents" (Andersen 1979 : 42). The "extended realignment" of 1928–36, however, was also closely associated with the impact of the Great Depression. Without this the new voters would have stayed home on election day, as did significant numbers of all voters throughout the period 1896–1928 (Burnham 1982 : 172–3). That period was marked by "party decomposition" and "vanishing turnout" as much as by regional party systems.

The "New Deal" system proper prevailed only until 1948. Under this system there was some nationalization of party organization and voting alignments. In particular, sectionalism was slowly replaced by the emergence of two-party competition where it had not prevailed for decades. Of greatest importance, the Democratic party became competitive in the Northeast, especially in the cities. The Republican party made somewhat slower inroads in the South. But, overall the politics of sectional dominance was disintegrating.

Since 1948–50, however, the New Deal system has itself been in retreat. Not so much through a revival of sectionalism, although this still exists, as through a tremendous increase in ticket splitting and decrease in partisan identification (Burnham 1982 : 112–13). Perhaps the Eighty-ninth Congress (1965–6) marked the final demise of the New Deal system. The enactment of major civil rights legislation finally broke up the alliance between the northern urban and the southern "plantation" wings of the Democratic party, although this coalition had been under stress ever since southern rebels ran their own "Dixiecrat" candidate for president in 1948 (Fig. 10.2). Sectional volatility rather than sectional dominance has been the overall result. Today, for example, the geographies of congressional and presidential elections frequently diverge (Bensel 1984: 386). Sectionalism remains much stronger in presidential than in congressional elections. At the same time, the sectionalism exhibited is markedly different between elections. Congressional elections are now more often *local* contests about national and local issues, whereas presidential elections are still national contests that tap regional and local interests and sentiments but with unpredictable *sectional* results (pp. 385–8). Moreover, since the 1960s, turnouts have once again been in decline. Though it was generally much easier to vote in 1980 than it was in 1960, especially for blacks and poor whites in the South, congressional turnout outside the South was 16.1 percent below 1960 levels and in the South was nowhere near previously all-time high levels (Burnham 1982: 181–3). It is instructive to note, however, and a measure of the nationalization of low turnouts, that for the first time in history Mississippi had a higher congressional turnout in 1980 than did New York (47.7 percent to 43.4 percent). This turnout issue will be addressed more centrally in Chapter 12.

Schattschneider (1960: 87) has described the 1932 election as marking the substitution of a "national political alignment for an extreme sectional alignment everywhere . . . except in the South." But over a period of years, and beginning before Schattschneider wrote, the ties that held the bipolar New Deal coalition together and discouraged the "old" sectionalism have broken. Two of the most recent "third-party" presidential campaigns represent the two most important centrifugal forces within the Democratic coalition. The George Wallace vote of 1968 indicated the southern dissatisfaction with the pro-labor and civil rights policies of the northern wing of the Democratic party. The John Anderson vote in 1980 represented the frustrations of certain elements in industrial, educational, and commercial electorates with Jimmy Carter's attempt to revive a "conservative" Democratic consensus.

The "new" sectionalism, however, is best seen in terms of Republican attempts to exploit the problems of the Democratic party. The recent strategy of the Republican party has been to go after the disaffected southern (and other) Democrats, and in so doing reverse the geographical pattern that was a feature of the system of 1896.

The strategy has a deeper logic, however. Since the late 1940s, and more especially since the late 1960s, rates of economic and population growth have been greater in the South and the West than in the Northeast of the United States. In a time of American decline within the world-economy (Agnew 1987a), this suggests that, in Burnham's (1981a: 380) words,

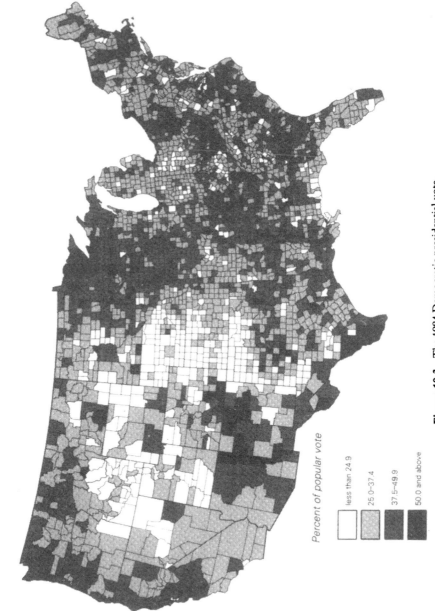

Percent of popular vote

less than 24.9

25.0–37.4

37.5–49.9

50.0 and above

Figure 10.3 The 1984 Democratic presidential vote.
Source: Archer et al. (1985 : 280).

Regional battles are likely to become major features of national political conflict, inside and outside Congress, irrespective of which president or party is in power. But under a conservative administration, the decline of the old industrial heartland will become more and more explicit. In particular, the vast increase in military consumption expenditures which is projected through mid-decade will considerably accelerate the transfer of wealth and power toward the South and West.

If this does indeed occur, the end of a 50-year absence of extreme sectional stress in the party system will also witness a reversal in party roles. The Republicans, the party of liberal capitalism and the former party of the American industrial core (the Northeast), would become the party of the periphery (the South and West). The Democrats would take over from the Republicans in the Northeast and, because of population shifts, probably become the minority party.

As yet, however, this has not definitively happened. The results of the 1984 presidential election do suggest relative Democratic strength in the Northeast, Republican strength in the West, competition in the South (Archer *et al.* 1985). But there is immense *local* variety (Fig. 10.3). For example, suburban areas were strongly Republican whatever the region, college towns and mining areas were strongly Democratic. This variety reflects in part the greater geographical *patchwork* of economic activities and cultural outlooks today than prevailed earlier in this century (Agnew 1987a). When this is added to the fading of the dominant regional orders (a cotton-based agrarian economy in the South, etc.), one might expect *increased* local political expression rather than a return of sectional politics or a revival of nationalization (Phillips 1978).

As Gramsci (1971 : 276) once observed "the crisis consists precisely in the fact that the old is dying and the new cannot be born; in this interregnum a great variety of morbid symptoms occurs." Certainly, there is little sign on the political horizon of that class politics which was effectively written out of American history as early as the 1880s. Congressional elections have become especially disaggregated: partisan swing is more and more a local phenomenon, and the electoral advantages of incumbency are increasingly rewarded in a politics in which parties have been displaced by what Blumenthal (1982 : 17–26) calls the "permanent campaign" of individual candidates. This is an electoral politics that is structured by the mass media, the candidates, pollsters, consultants, and advertising agencies rather than by parties (see Ch. 12). Under present institutional arrangements, therefore, the prospects are for a continued interregnum – volatility, declining participation in elections, and the continued decay of the major parties. The most interesting correlate of all this is that it seems very much the case, as asserted by "Tip" O'Neill, the recent Speaker of the US House, that, in the United States today, "All politics is local."

Table 10.1 The "bipolar structure" of American politics: trade area alignments on high-stress roll calls (%)★ with the "industrial–commer-cial core," 1885–1976.

Trade area	1885–7	1895–7	1905–7	1915–17	1925–7	1935–6	1945–6	1955–6	1965–6	1975–6	100 roll calls 1885–1976
Philadelphia	100	80	90	100	100	100	95	100	95	100	96.0
Boston	100	65	80	100	100	100	70	100	95	100	91.0
Buffalo	100	70	85	100	95	–	90	90	85	65	86.7†
Chicago	90	80	100	100	80	60	45	85	90	80	81.0
New York	25	50	90	100	100	70	60	100	100	100	79.5
Detroit	20	65	90	80	90	90	80	70	90	90	76.5
Pittsburgh	100	80	100	100	100	10	75	60	85	55	76.5
San Francisco	100	50	60	100	85	35	40	90	65	95	72.0
Cleveland	75	90	100	55	50	75	40	90	50	70	69.5
Omaha	100	75	100	65	25	–	90	50	45	–	68.7†
Cincinnati	75	80	80	60	85	95	85	60	40	20	68.0
Minneapolis	100	75	100	90	80	45	85	20	35	40	67.0
Indianapolis	35	90	–	–	25	50	100	60	40	75	59.4†
Kansas City	90	50	90	0	45	60	75	25	30	20	48.5
Denver	–	15	–	25	80	10	85	30	60	20	40.6†
St. Louis	0	80	30	0	15	40	85	40	60	25	37.5
Baltimore	10	60	15	25	25	30	20	85	–	–	33.7†

Louisville	15	25	0	0	90	60	–	–	25	10	28.1†
Richmond	0	0	0	0	20	40	0	30	20	10	12.0
New Orleans	0	5	0	0	20	10	0	10	30	0	7.5
Birmingham	–	–	0	0	0	20	0	0	25	0	5.6†
Atlanta	0	0	0	0	0	15	5	10	10	0	4.0
Memphis	0	0	0	0	0	10	0	0	20	0	3.0
Dallas	–	–	–	0	5	15	0	0	0	0	2.5†
Correlation**	0.729	0.760	0.892	0.898	0.762	0.668	0.725	0.810	0.821	0.865	

* Where a trade area split evenly on a high–stress roll call, the delegation was viewed as half "in–agreement" with and half "opposed" to the polar position. Otherwise, the trade area majority determined the position of the delegation.

† Table 10.1 does not report the sectional alignment of trade areas that did not appear consistently in trade area systems between 1880 and 1981, nor does the table record the position of delegations containing fewer than four members. In eight of the twenty–four delegations, the trade area either did not exist as a separate entity or had fewer than four members in one or more of the ten Congresses. In these eight cases, the performance of the trade area delegation in the remaining Congresses was pro–rated to one hundred.

** Relationship between roll–call behavior in the indicated Congress and behavior in the nine other periods (the summary scores were recalculated to exclude the indicated Congress).

Source: Bensel (1984 : 36).

CONTEXTS OF ELECTORAL BEHAVIOR

Clearly, one of the most distinctive features of American politics since the Civil War has been its local and regional nature. Bensel (1984), following in the footsteps of F. J. Turner (1932), J. Turner (1970), and V. O. Key (1964) argues that this geographical differentiation is grounded in an interregional division of labor that has continually produced a struggle for control over the national political economy. Different parts of the country have had distinctive and different interests that could be furthered or hindered by various national policies. Consequently, control over or considerable influence within the federal government becomes a primary goal of local representatives.

Bensel demonstrates his argument by examining the roll-call votes of US congressmen on "critical" policy issues in ten separate congresses in the middle years of each decade from 1885–7 to 1975–6. The issues include union veteran pension bills for 1885–7, public works bills for 1935–7, revenue sharing for 1975–6, etc. Congressional districts are grouped by "urban trade areas." These are hinterlands measured by a number of central-place criteria such as rail network areas (1881–91), Federal Reserve banking centers (1901–21), wholesale trade distribution areas (1921–41), or Rand McNally marketing divisions (1941–81). Indices for "sectional stress" on individual roll calls are calculated. These measure the cohesion of trade area congressional delegations relative to random voting in order to estimate "sectional alignments" on critical issues. These are averaged across all ten congresses to give a stress score for each trade area and correlated across trade areas for each congress to measure the degree of sectional stress over time (Table 10.1).

The picture that emerges is of persisting if cyclical sectional alignments. The summary scores identify what Bensel calls an "axis of sectional stress" running between the Northeast and the Deep South. The correlations reveal the persistence or decline of stress. Of the ten congresses: only three (1885–7, 1935–6, and 1945–6) stray beneath a correlation of 0.75. The lowest, that of 1935–6, represents the "flood tide of the New Deal" (0.668). Since then sectionalism by trade area delegations has returned to earlier levels. This is particularly marked at the "poles" or ends of the sectional continuum. Thus, Philadelphia, which supported the "core position" on 96 of 100 high-stress votes, comes out at one end; Memphis, which opposed the "core position" on 97 of the same roll-call votes comes next to last at the other end. A "swing group" of trade areas, mainly in the West, lie more or less consistently between the northeastern "core" bloc and the southern "periphery" bloc.

Bensel (1984: 38) explains the cyclical pattern of sectional stress as follows:

The passing of Reconstruction, the consolidation of local power by the southern plantation elite, and tariff-protected industrialization all combined to intensify and mold political conflict along a single sectional axis. As the core and periphery elites increased their dominance over the

Table 10.2 Average sectional stress and distribution of scores on competitive roll calls, 1885–1976.

| Congress | Average sectional stress | Percentage of competitive roll calls in which sectional stress index was: | | | | | | N |
		Over 70.0	60–69.9	50–59.9	40–49.9	30–39.9	20–29.9	
49th (1885–7)	55.0	0	40.6	34.3	14.1	9.4	1.6	64
54th (1895–7)	48.5	0	12.5	33.3	41.7	12.5	0	24
59th (1905–7)	58.8	0	47.6	42.9	4.8	4.8	0	21
64th (1915–17)	54.9	3.6	21.4	53.6	21.4	0	0	28
69th (1925–7)	55.1	0	20.0	53.3	26.7	0	0	15
74th (1935–6)	44.5	0	0	27.8	33.3	33.3	5.6	18
79th (1945–6)	48.1	0	2.2	45.7	39.1	13.0	0	46
84th (1955–6)	47.5	2.9	2.9	28.6	40.0	25.7	0	35
89th (1965–6)	39.4	0	0	5.7	39.6	45.3	9.4	53
94th (1975–6)	39.9	0	.6	7.1	41.0	44.2	7.1	156

Source: Bensel (1984 : 53).

industrial-commercial and interior-hinterland poles, the correlation between the policy conflicts associated with each period and the underlying structure of sectional stress steadily increased from 1885 to 1917. Later the farm crisis of the twenties and the New Deal introduced "cross-cutting" cleavages in regional conflict, and the correlations fell between 1917 and 1936. From the point of greatest deviance in the 1930s, the correlation at each subsequent period increased.

But it is important to note that the *intensity* of *sectional* stress *has* declined over time. Bensel's data in general and his maps showing roll-call votes on specific issues in particular are indicative of *decreasing* regional polarization (Table 10.2). This does not necessarily contradict the findings summarized in Table 10.1. It suggests that polarization has become more local and unpredictable rather than regional and certain. Bensel perhaps overplays the persisting regional element of polarization in Table 10.1 at the expense of noting the increased volatility and individuality of *trade-area delegation* voting behavior. Local rather than regional economic interests are still clearly at issue, but at a time when the *identity* between them has been much reduced.

Bensel (1984) contends that recent institutional–political changes presage a resurgence of classic American sectionalism. One of these is the decline of the congressional committee system. A strong committee system, he argues, "allowed the representatives of core and periphery regions to carve up the national political economy into sectors which the sectional poles could then control" (p. 53). For example, the periphery controlled some committees – defense, agriculture, public works – and the core controlled others –

education, labor, and banking. This system both removed decisions off the floor of the House and into private rooms where deals could be struck, and produced ambiguous and discretionary policies that could allow for reciprocal or log-rolling votes. The recent decline of this system, under pressure for democratization of House rules and modifications of the seniority preference, has removed a major force for vote trading and thus led to increased sectional stress.

A second push for renewed sectionalism is said to come from the emergence of regional "ideological protocoalitions" which represent regions *across* party lines. In particular, "liberalism" is now defined in the United States in terms of the political imperatives of the industrial core. Consequently, representatives from this region form a sectional bloc irrespective of party.

Whatever the merits of Bensel's case, and he makes a most persuasive one, his single-minded focus on *sectionalism* obscures a most important part of his own argument. The building blocks of his empirical analysis are not sections or regions as such but much smaller trade areas. His own case rests on the idea that sectional patterns are aggregates of local processes defined by the economic–political interests of local populations. Even as sectional patterns have waxed and waned, therefore, place-specific estimations of local advantage and disadvantage have been at the root of the shifts.

Today, the great sections of the United States are much less homogeneous economically and socially than they were in the past. Given patterns of plant closings and openings and job losses and gains over the past twenty years, it is especially apparent that the "industrial core" has lost its coherent geographical form (Bluestone & Harrison 1982, Agnew 1987a). Patterns of economic growth and decline are now clearly "patchy" and local rather than sectional in the orthodox sense.

From World War II until the late 1960s, US industry decentralized from the confines of the "industrial core" or manufacturing belt. This process had begun during World War II when the War Production Board had as one of its objectives dispersal of new industrial capacity to "uncongested areas." This also stimulated population movement to the places where the new industries were located. This movement was mainly to the West, but by the 1960s had tilted towards the South (Table 10.3). The relocation policy thus had a built-in multiplier effect. As most of the new industrial capacity developed during the war was "defense"-related, the continued military build-up after the war mainly benefited the places where it was already located.

The 1950s and 1960s were decades of explosive growth in the US economy as a whole. Involved in this expansion were two important changes that had major geographical implications: the development of new transport and communication technologies, and a continuing centralization of capital that provided the means to exploit the new technologies. The construction of the interstate highway system, the development of

Table 10.3 Yearly percentage of population growth by region, 1940–80.

Region	1940–50	1950–60	1960–70	1970–80
Northeast	1.09	1.56	1.01	−0.01
North Central	1.56	1.76	1.06	0.33
South	2.61	2.59	1.38	1.75
West	3.75	3.50	2.37	2.08
United States	2.14	2.26	1.36	1.02

Source: Leven (1981).

computers and long-distance communication systems, and the growth of air travel provided a permissive technological environment in which businesses could decentralize their manufacturing operations yet maintain central control. In particular, these technologies made it possible for businesses to take advantage of new sources of cheaper and more tractable labor in "peripheral" locations in the US (and elsewhere).

The development and use of these new technologies were stimulated by the large firms that came to control significant chunks of American industrial capacity after World War II (Table 10.4). Firms that can both dictate price, largely free of competitive pressure, and restrict output by barring the entry of new firms can earn monopoly profits, control technological change, and exercise considerable political power. More important in the context of the present discussion, dominant firms can switch their capital resources geographically without competitive pressures to constrain their decisions.

Technological change and corporate concentration were therefore complementary. The new technologies allowed the new super-firms to fashion capital-switching and relocation strategies without giving up centralized control. In these circumstances, locations outside the historic industrial core could become competitive. The new technologies also made unskilled labor, as long as it was cheap and nonunionized, more rather than less attractive. The highly skilled labor to be found largely in the historic core was no longer needed in many industries.

The "aging" of the industrial core was also involved. Old industrial plant in industries with declining profit rates relative to newer industries presented a barrier to many firms' future growth. Indeed, many of the presumed advantages of the industrial core in terms of infrastructure and multiplier effects became disadvantages. In particular, high levels of unionization retarded the introduction of new technologies and the costs of maintaining aging public facilities and New Deal social welfare programs (and taxes) were ones that businesses were anxious to avoid.

Federal government policies and spending were also important in encouraging shifts in investment from the industrial core to places outside. The "defense" industry has been noted previously. But government agricultural,

Table 10.4 Percentage of sales accounted for by the four largest producers in selected manufacturing industries, 1947–72.

Industry code	Industry	Percentage of sales	
		1947	1972
2043	cereal breakfast foods	79	90
2065	confectionery products	17	32
2067	chewing gum	70	87
2092	malt beverages	21	52
2211	weaving mills, cotton	18[a]	31
2254	knit underwear mills	21	46
2279	carpets and rugs	32[b]	78
2311	men's and boys' suits and coats	9	19
2337	women's and misses' suits and coats	3[a]	13
2421	sawmills and planing mills	11[a]	18
2771	greeting card publishing	39	70
2829	synthetic rubber	53[a]	62
3211	flat glass	90[a]	93
3312	blast furnaces and steel mills	50	45
3421	cutlery	41	55
3511	turbines and turbine generators	90[b]	93
3555	printing trades machinery	31	42
3585	refrigeration and heating equipment	25	40
3624	carbon and graphite products	87	80
3633	household laundry equipment	40	83
3636	sewing machines	77	84
3641	electric lamps	92	90
3661	telephone and telegraph apparatus	90	94[c]
3674	semiconductors and related devices	46[b]	57
3679	electronic components	13	36
3692	primary batteries, dry and wet	76	92
3711	motor vehicles and car bodies	92[d]	93
3724	aircraft engines and engine parts	72	77
3743	locomotives and parts	91	97[d]
3861	photographic equipment and supplies	61	74
3996	hard surface floor covering	80	91

[a] 1954. [b] 1963. [c] 1970. [d] 1967.
Source: Bluestone & Harrison (1982 : 120).

public works, and general fiscal policies have also stimulated diffusion of industrial activities in the United States. For example, federal income tax policies encourage investment in new facilities at the expense of upgrading existing ones (Luger 1984).

Since the late 1960s the process of industrial decentralization has accelerated. By 1980 there were as many manufacturing jobs outside the core as inside it (Agnew 1987a, Ch. 4). This was largely the result of the changing international position of the US economy. The regions most vulnerable to foreign competition – the East North-Central region (including Chicago and

Detroit) and the Northeast – lost jobs, have declining rates of capital investment, and have higher unemployment rates. Other regions have weathered the storm more successfully, partly because of their more attractive "business climates" and partly because of their lower energy costs.

But the redistribution of industrial activity and concomitant shifts in population and service-sector activity have not been at a regional scale, as such. Rather, only parts of the core have gone into long-term industrial decline, and only scattered places in other parts of the country have benefited from recent economic changes. The East North-Central or Great Lakes region with its high proportion of employment in durable goods manufacturing has been most exposed to changes in the world-economy without the compensation of high levels of defense or other federal spending (Browne 1983, Cloos & Cummins 1984). Other parts of the industrial core, especially some places in New England, have rebounded because of increases in defense spending since 1978, high levels of foreign direct investment, and lower relative wage rates (Browne & Hekman 1981, Little 1985, Browne 1984).

A number of cities and associated hinterlands outside the core now display features akin to those characteristic of the "classical" core. For example, San Francisco shares all the major features of leading northeastern centers such as New York and Boston. Places such as Atlanta, Dallas, Houston, Los Angeles, and Miami are on the same track. At the same time, other places are economically stagnant or declining. The division of the United States into two large and mutually exclusive economic regions around which political conflict completely revolves is therefore no longer, if it ever was, an accurate representation of an essentially locality-based pattern of economic growth and political expression.

Bensel's own data in fact support much of the foregoing argument. However, it may be useful to point to an interesting example of the importance of place in structuring political expression from an older era of American history when sectionalism was a more powerful phenomenon than it is today. This is the case of socialism in America and what Bennett and Earle (1983) call "a geographical interpretation of the failure." After reviewing the two major explanations for the failure of American socialism, the prosperity of American workers and the tactical failures of leaders and trade unionists, Bennett and Earle trace the support for Socialist candidates, trade union membership, strike activity, and local modes of economic activity between 1865 and 1920. They find that socialism failed to "take off" not so much because of a general prosperity of American workers but because of a deep division in the material interests of skilled and unskilled workers. This division was deepest in the larger cities, with their greater industrial diversification and more marked wage divergence between skilled and unskilled. Class politics did take root, however, in some smaller cities in the Upper Midwest, Ohio, Western Pennsylvania, and around Albany, New York. But these were places experiencing a "crisis of transition" from a preindustrial to an industrial mode of economic activity. Once they had traversed this phase they too came to share the social divisions of the larger cities – and the drift away from socialism.

Katznelson (1981) argues, however, that over the long term and beginning in the 19th century, American politics has been based on *values* and *practices* that have been inimical to the development of "class consciousness" and socialism. He isolates the varied ethnic backgrounds of American immigrants as of special importance. "Torn by the traumas of migration and proletarianization, ethnic and racial groups have tended to enclose themselves within spatially as well as culturally specific communities" (p. 10). Consequently, the politics of community became separated from the politics of work. Struggles in the workplace and trade union activities were separated from the "regular" politics of the neighborhood and the ballot box. The system of organizing elections on territorial lines put a premium on "the identities that are to be found where people live rather than where they work" (p. 71). The outcome has been the impossibility of channeling whatever class consciousness there might be in the workplace into the political system at large. Katznelson concludes (p. 72):

> American socialists achieved most when one of two very special conditions obtained: when workplaces and residence communities were tightly bound together as in the Finnish mining communities of Northern Michigan; or when the voting public consisted principally of first-generation immigrants from Europe who brought a socialist tradition with them. In almost all other cases, the city trenches held against Socialist incursions.

If one bias in Bensel's interpretation of the geography of American politics then is the persistent reading of a sectional pattern from a much "spottier" map, another is the neglect of sources of geographical variety in political expression other than the sectional–economic. It is a commonplace of American electoral studies that economic concerns *do* occupy a central place in American political life. Dozens of public opinion polls have shown economic problems to be *the* major source of public concern (see Kiewiet 1983). This is not surprising given the dominance of business in national life and national agenda setting. But few observers deny that this concern is not grounded in a frame of reference that is as much cultural and moral as it is economic. Consequently, political expression cannot be simply "read off" from *current* economic conditions. Current economic conditions are given meaning and produce political behavior through the "filter" of historical–cultural context.

Elazar (1972), for example, provides a threefold typology of "founding" cultural orientations out of which more recent place-specific ones have been blended. These "embrace and shape the primary social, economic, and psychological thrusts that influence American politics" (p. 85). The three "political cultures" – the "moralistic," the "individualistic," and the "traditional" – are viewed as having their origins in the New England, Middle Atlantic, and Old South regions of original settlement. As they spread across the continent from their Atlantic origins they were altered, amended, and combined in different ways. The final result is a mixed pattern of political cultures across the country because, as Elazar (1972:89) puts it, they are "rooted in the cumulative historical experiences of particular groups of people." And, one must add, in particular places.

The "traditionalistic culture" reflects a precommercial attitude that accepts both hierarchical social order and personal relationships between patrons and clients rather than impersonal or bureaucratic ones. The "individualistic culture" emphasizes the marketplace as the central image of the "good society." It thus follows an emphasis on the centrality of private concerns and the need to "limit" government lest it interfere with the natural order of the market. This is pre-eminently the ideology of the world-economy and the one that has predominated nationally in the United States. The "moralistic culture" sees commitment to community and public life as morally superior to the magic of the market place. By virtue of its public orientation, this culture produces a greater commitment than the others to active government intervention in economic and social life. But it is not collectivistic. At root it rests on making moral appeals to the individual conscience. It is not socialism, American style (Baltzell 1979).

These original "founding" cultures reflect both the ideas and practices brought by founder groups and beliefs that proved functional in the context of settling a "new land." The traditionalistic culture was the product of both transplanted "establishment" attitudes from rural England and the realities of a slave-based plantation economy. The individualistic culture emerged from the origins of the Middle Colonies as a region of competitive commercial agriculture. The New England moralistic culture was a product of the Puritan ideals of its first settlers and their communal pattern of settlement.

Each of these cultures was in a fundamental sense incompatible with the others. But each was also altered as it came into contact with the others. Thus today, although there is a latitudinal bias corresponding to the locations of the original source-areas, and the individualistic predominates nearly everywhere, the three cultures have "melded together" into particular combinations in different places, reflecting patterns of migration and historical experiences. The moralistic and individualistic predominate in the Northeast, Upper Midwest, and Pacific Northwest, the traditional and the individualistic in the South and Lower Midwest, and a variety of combinations prevail in the Southwest, the Mountain States, and California. Southern California is unique in combining all three cultures in volatile combinations (Elazar 1972).

It is out of the generally volatile West that many of America's "third party" and other political revolts have come. For example, the recent New Right takeover of the Republican party involved the orchestration of a variety of social issues originally brought into the public eye through the referendum process available to monied and disaffected groups in the West (Wolfe 1981). But the West also provides an unstable and uncertain cultural environment in which people can be easily persuaded that this or that issue – Cuba, women's rights, abortion, gun control, or the Panama Canal Treaty – threatens the persistence of their social worlds and the values and interests they hold dear (Davis 1981). Contemporary economic conditions are likewise viewed through the lens of cultural meaning as well as in terms of the immediate material interests they elicit. They cannot be understood satisfactorily just in economic stimulus–political response terms.

There are also important religious and ethnic cleavages in the United States that are interwoven with the basic material–cultural ones. As Burnham (1981b : 32) suggests when commenting on the role of religion in the 1980 presidential election, "the United States is, certainly statistically, God's country to a quite remarkable extent." Although modernization theories see secularism as a major feature of "advanced industrial societies" such as the United States, this is in fact far from the case. At least 40 percent of probable voters in 1980 reported having had a personal experience with Jesus Christ (Burnham 1981b). All the major presidential candidates that year – Carter, Reagan, and Anderson – professed to having been "born again" at one time or another. But, as Burnham (1981b) notes, the United States is exceptional among the more "developed" societies in the extent of its professed religiosity. This owes something, he suggests, to the dominant myths and symbols established by the founders and to the American–Soviet rivalry in the contemporary world, where religion is favored as a direct popular reaction to the official atheism of the Soviet regime. However, rather than a permanent affair, the direct influence of religion in American politics has come in waves, such as during the millenarian revival of the period 1840–60 in New England and Upstate New York; the nativist (and Protestant) ferment over Catholic immigration and Catholics in politics in 1844–60, the 1890s, 1918–30 and 1960; Prohibitionism (1850s, 1880s, 1918–30); the spread of the Ku Klux Klan outside of the South (1918–25); and the rise of the moral majority and antiabortion groups in the late 1970s and early 1980s. Such episodes have varied as to their causes, but generally correlate with periods of rapid economic and social change. Of course they also correlate with the locations of the most threatened and most active elements mobilized to meet the threat. They are also often denominational in nature. Evangelical or Fundamentalist Protestants have, at least until recently, tended to be those most likely to self-consciously introduce religious issues into the political arena. Perhaps their close association with the founding groups legitimizes this intervention in their own, and other's, eyes.

But religion, in large part as a badge for ethnicity, has wider influences on political behavior. The original and older immigrant groups, with the exception of the Irish, tended to be Protestants of one type or another. Later immigrants, however, were Catholic or Jewish. As a consequence the South is still largely Protestant, having received relatively few of the later immigrants, and large parts of the Northeast are predominantly Catholic. Other areas are mixed or contain "outliers" or islands where one denomination or religious affiliation is predominant (Shortridge 1976). Most of the Jewish population of the United States lives in the New York metropolitan area, or in other large cities such as Chicago, Los Angeles, Miami, Philadelphia, and Atlanta. Certain unique religious groups are concentrated in a specific "homeland" (such as the Mormons in Utah) with a strong representation also in outliers in neighboring states.

At one time there was a strong association outside the South between Protestantism and Republicanism, and between Catholicism and Judaism and support for the Democratic party. But these associations always varied considerably by ethnicity and from place to place. For example, Italian Catholics were often more inclined to be Republicans than were Irish Catholics, and all voters in Upstate New York and downstate Illinois were more inclined to vote Republican than their coreligionists in the "big cities" of New York and Chicago, respectively. The association between religious *affiliation*, if not practice, and political behavior is relatively muted today, even if the *general* relationship persists (Key 1964, Asher 1980, White 1983).

However, ethnicity is usually mediated through place in the form of the segregation and the reinforcement of group identity this provides. Today two groups are most important in this respect: blacks and Hispanic-Americans. Blacks were, of course, formerly concentrated in the South, but today they are also to be found in dense clusters or ghettos in most of the major and many of the smaller cities elsewhere. They now constitute a major bloc of support for the Democratic party, but low turnouts and general political disaffection reduce their political impact. Hispanic-Americans, mainly Mexicans and Puerto Ricans, are likewise geographically concentrated. Most Mexican-Americans are in Texas, New Mexico, Arizona, and California. Most Puerto Ricans are in New York City. Cuban-Americans are a major presence, socially and politically, in south Florida. The Mexican-Americans and Puerto Ricans have low levels of political participation but tend towards Democratic party candidates in the places they live. Many Cuban-Americans are, to the contrary and because of their status as middle-class refugees from Communist Cuba, supporters of the Republican party in national politics.

The immense and geographically uneven religious–ethnic variety of the United States, however, is not alone in reinforcing the material–cultural basis for place-specific political behavior. A major feature of American social–political organization is the split within urban areas between central cities and their suburbs in the provision of an array of public goods and services. By way of example, just as in Scotland (see Ch. 7) the Labour party built its popular base on "municipal socialism", so the northern Democratic party in the United States built its popular base on patronage and provision for immigrant and minority groups. The New Deal and the Great Society (of the late 1960s) illustrate two features of this process. First and most importantly, national Democratic leaders sought to ensure that a mobilized urban electorate would identify with the Democrats and thus offset the Republican advantage elsewhere. Only by building sufficiently large urban majorities could they hope to win national elections (Andersen 1979). Secondly, however, new local programs were constantly necessary to attract new urban constituencies. This put tremendous strain on local fiscal resources at the same time that it encouraged the creation of local "growth coalitions" to enhance or reinforce the cities' economic bases (Mollenkopf 1983).

Simultaneously, however, a massive population redistribution was going on within urban areas. People from both core cities and rural areas have poured into suburban rings, shrinking the city's share of metropolitan population. Large cities reached their peak populations around 1930 and have been declining ever since. Suburbanization has a variety of roots, but its pace is not unrelated to the importance given to housing and real estate as economic stimulators by post-World War II American federal governments (Walker 1978). Its appeal lies partly in the particular combinations of public services and private goods that different suburban areas provide, and the protection they afford from the costs of helping meet the agenda of the urban Democrats.

Increasingly, however, central cities and suburbs are also in competition with one another for economic growth (Logan 1978, Cox 1984). A major form of political activity has become that of harnessing local government to compete against other places for economic growth. This usually involves attempts to finance business services through public funds, and enhance business climates through tax abatements and rezonings. Inevitably, suburbs have all sorts of advantages over central cities in this process. Their very basis in a *limited* conception of government as a provider of services and a lack of a vibrant public life make them more attractive to business than the more overtly political and conflict-ridden central cities. Although it should be noted that suburbs differ greatly in the degree to which they are involved in "place competition." By definition, "industrial" suburbs are much more involved in competition for industries than are upper middle-class residential suburbs, though all are involved in the struggle for the appropriation of wealth and the minimizing of social costs (Molotch 1976; Logan 1976a,b).

But just as there is a variety of central city forms, with Republicans doing remarkably well in some, so there is a remarkable diversity of suburban forms. Though it is true, as noted by Archer *et al.* (1985), that suburban areas throughout the United States voted predominantly Republican in the 1984 presidential election, the stereotype of a Republican suburb was produced initially by the successes of Eisenhower in the 1952 and 1956 elections rather than any longer-term trend. In particular, there is not a single suburban type, but numerous types of suburb (Wirt et al. 1972).

What is clear, however, is that urban life in America has become more, not less, segregated over time. Population statistics show that urban America now has especially marked racial "ghettos" in its large central cities, surrounded by largely white, if not always affluent, suburbs. As the exodus to the suburbs and to small towns beyond continues, cities are increasingly losing their middle-class, white, higher-income, tax-paying populations. Increasingly, nonwhite, lower-income, low-education, unskilled, and "marginal" populations are being concentrated in central cities. In these circumstances, and to the extent that the Democratic party continues to be identified with the central cities and their populations, the suburbs, whatever their particular attributes, will probably become increasingly Republican. It is indeed the American suburbs that will have saved the Republican party from the slow death that was regularly predicted for it in the 1960s and 1970s (Asher 1980).

Conclusion

Rather than disappearing under the irresistible pressure of nationalization, therefore, place survives in the United States both in providing the fundamental setting for political behavior and in the form of distinctive political expression. Much as geological strata represent past processes existing into the present, so places are built up out of the interpellation of past material–cultural practices with present-day ones. Places thus mediate between political organizations/institutions and political behavior not by the immediate impacts of sectional–regional economic processes or national social cleavages but through the ways these impacts are interpreted and given meaning by local populations engaging in their regular routines and interactions. Though in the past such processes were often more similar across places and one could speak readily of sectionalism, and for a while of "nationalization," today the decomposition of the major political parties and the increased "patchiness" of economic development and cultural outlooks across state and metropolitan areas point explicitly to the continuing importance of place in the structuring of American political life.

11 *Place and American politics*
The historical constitution of political behavior in places

The major problem with simply examining and attempting to explain political behavior at an aggregate level is that the complex interactions of place and politics are reduced to generalities. There is nothing inherently wrong with this. It is certainly preferable to generalizing from the particular. The danger in selecting specific places for closer examination is that they will be seen solely as case studies or typifications of regions or types of place. But there are more specific rewards to study of the variety of human experience and political behavior in particular places. These are those eloquently expressed by Eudora Welty, the writer of Jackson, Mississippi, in *The eye of the story* (as quoted in Kirby 1984: 179),

> Place absorbs our earliest notice and attention, it bestows on us our original awareness; and our critical powers spring up from the study of it and growth of experience inside it. It perseveres in bringing us back to earth when we fly too high. It never really stops informing us, for it is forever astir, alive, changing, reflecting, like the mind of man itself. One place comprehended can make us understand other places better. Sense of place gives equilibrium; extended, it is sense of direction too.

From this point of view, the outsider can never capture the intricacies of a place like an insider. The observer, as opposed to the participant–observer, must be satisfied with conveying something of the life of a place without capturing its most intimate details. Ultimately, this is something resident novelists often do best.

As in Chapter 7, four places are the focus of this chapter. They are Detroit, Michigan; Mansfield, Ohio; Middlesboro, Kentucky/Claiborne County, Tennessee, and Miami, Florida. They are selected, as were the places in Chapter 7, in an attempt to convey the rich variety of experience in different places. But they also represent a range of relationships with wider economic and political worlds that should become apparent as each place is described.

Table 11.1 Population growth of Detroit, 1830–1980.

Year	Total Population	Total % change	Black Population	Black % change	Black % of total population
1830	2,200	–	73	–	3.3
1840	9,700	340.9	121	65.8	1.2
1850	21,019	116.7	587	385.1	2.8
1860	45,619	117.0	1,403	139.0	3.1
1870	79,517	74.3	2,231	59.0	2.8
1880	116,340	46.3	2,821	26.4	2.4
1890	205,853	76.9	3,431	21.6	1.7
1900	285,704	38.8	4,111	19.8	1.4
1910	465,766	63.0	5,741	39.6	1.2
1920	993,678	113.3	40,838	611.3	4.1
1930	1,568,662	57.9	120,066	194.0	7.7
1940	1,623,452	3.5	149,119	24.2	9.2
1950	1,849,568	13.9	300,506	101.5	16.2
1960	1,670,144	−9.7	482,229	60.5	28.9
1970	1,511,482	−9.5	660,428	37.0	43.7
1980	1,203,339	−20.4	758,939	14.9	63.0

Source: US census data.

Detroit

The modern history of Detroit is in large part the history of the rise and decline of the US automobile industry. Detroit's growth as an automobile center began when Ransom E. Olds of Lansing, Michigan, established a factory in Detroit in 1899. Detroit proved to be a strategic location for this fledgling industry. It was halfway between the iron-ore fields of Minnesota and the coalfields of Appalachia, and accessible to each by cheap water transportation. Moreover, capital from Michigan's timber and mining industries financed a wide range of metal-working industries. Prime among these were the component and assembly plants of various auto manufacturers, notably Henry Ford.

Between 1910 and 1920, the major era of the city's growth, the population of Detroit increased by 113 percent (Table 11.1). Of the 993,000 inhabitants in 1920, 289,000 were recent immigrants. The Poles (56,624) constituted the largest ethnic group, followed by non-French Canadians (55,216), Germans (30,238), and Russians (27,178). There were 41,000 blacks and numerous second-generation immigrants, of whom the Irish and Germans were the most numerous (Glazer 1965).

Detroit's image as an "immigrant city" was well established. From 1880 until the 1920s the various ethnic groups were more or less segregated from one another in distinctive ethnic enclaves (Zunz 1982). Politics was likewise an ethnic affair. In particular, the Catholic immigrant groups were predominantly affiliated with the Democratic party, and the Protestant and older "nativist" elements were more Republican in affiliation. Zunz (1982:107) sees it as follows: "It was along ... dividing ethnoreligious political lines that most issues, especially temperance and observing the Sabbath, were fought in late 19th-century Detroit." The Republicans were the representatives of a "moralistic" political order, they were the "prohibition party"; against them the Democrats represented the saloon keepers and the liquor interests.

However, by 1900 these "older" issues were transcended by the problems of equipping an expanding city with public services and delivering these to different parts of the city. But neighborhood reigned supreme in determining social and political behavior. Even as the locations of factories changed, and in particular as industry decentralized from the central city, ethnic neighborhoods maintained their cohesion and importance (Zunz 1982:181–3). Residence was not determined by workplace. Poles focused their residence and energies on the church. Other groups had other loci.

If ethnic diversity divided Detroit into ethnic enclaves, industrialization unified the city into a whole. However, no single institution was able to cut through ethnic barriers and unify the city's population. For example, though inequality was widely visible, the labor movement was unable to create a persistent sense of working-class consciousness. Until the 1930s Detroit was one of the least unionized major cities in the United States (Zunz 1982:224). People relied on family and ethnic institutions rather than unions when times were bad or when faced by lay-offs.

With the phenomenal growth of the city between 1910 and 1920 Detroit "was a place where – more than any other place in the United States – the industrial society was changing the way people lived" (Zunz 1982:309). As the growth rested on the automobile, so did the life of the city come to depend on the ups and downs of the automobile industry and the political aspirations of its owners.

Since most laborers and many craftsmen in the automobile industry had no union or other organizations to represent their interests, the industrialists of Detroit had free rein to reorganize both factories and the physical structure of the city itself. Henry Ford led the way. His efforts at rationalizing factory work are well known. But his activities as a proponent of "Americanization" and "civic reform" (Zunz 1982:309–25) are less familiar. Ford, and other industrialists, aimed at putting "order" into their factories *and* the city. They tried to stabilize their workforces by promoting conformists and eliminating those who were not "thrifty and of good habits" (p. 312). Their efforts at control outside the factories concentrated upon "strengthening" the office of mayor and council through at-large elections and lessening the influence of the ward-based, immigrant, and working-class organizations. After the charter of 1918 went into effect, the city could be run as a corporation – without "partisan" politics (i.e., without the Democrats).

As Detroit was becoming the "motor city," ethnic segregation began to recede in importance. By 1920 occupational variety was no longer character- istic of the major ethnic concentrations. What Zunz calls "cohesive socio- ethnic neighborhoods" were replacing the older cross-class ethnic neigh- borhoods in both city and newer suburbs. There was one exception: the city's black population was becoming increasingly segregated from the white, "drawn into a ghetto solely on the basis of race and without regard to their social status" (Zunz 1982: 354). Blacks were seen as an economic and political threat, as witnessed to by the rise of the Ku Klux Klan to political prominence in the early 1920s.

These trends continued until World War II (DeVito 1979). However, the Great Depression of the 1930s had an immediate and heavy impact upon the automobile industry. This "repoliticized" the city, bringing a surge of political activity and leading to the creation and rapid growth of a union for the autoworkers, the Union of Automobile Workers (UAW). In particular, the Democratic party re-established itself in the city after the interregnum of the 1920s, and has remained the major political force in state and national elections ever since. The change was rapid. Whereas in 1930 the two congressional districts entirely in the city were both held by Republicans, with large majorities (82.2 percent and 78.8 percent of the vote, respectively), by 1932 the Democrats held the first of these seats, with 68.1 percent of the vote. In 1936 the Democrats took both, with 80.4 percent and 55.1 percent of the vote, respectively, and have held them ever since. Local politics remains officially "nonpartisan" (Banfield 1965: 55).

In the late 1930s, and until the 1960s, rearmament and the rebirth of the automobile industry led to large-scale industrial expansion. But much of the new construction and many of the new jobs were in suburban areas. This provided a major stimulus to suburbanization of the city's population – though other factors, such as new highway construction and federal pro- grams favoring purchase of new houses, were also involved (DeVito 1979). Since the 1950s, suburbanization, particularly of the city's white population, has continued apace.

Detroit is still synonymous with the US automobile industry. For many journalists, "Detroit" is indeed shorthand for that industry. But in the past 25 years Detroit has lost 27 percent of its population and 70 percent of its jobs in manufacturing. By 1981, over 400,000 Detroit residents, one in every three, was receiving some form of public assistance (Luria & Russell 1984). Detroit's past growth was set by the cycle of auto production. Today the global reorganization of the automobile industry places the "motor city" at the heart of the crisis facing American manufacturing industry and the communities in which it is located.

In 1980–1 the US auto industry experienced its worst downturn since the

Table 11.2 Sources of components for Ford's "World Car": the European Escort.

Country	Components
Austria	radiator and heater hoses, tires
Belgium	hood-in trim, seal pads, tires, brake tubes
Canada	glass, radios
Denmark	fan belts
France	seal pads, sealers, tires, underbody coating, weatherstrips, seat frames, brake master cylinder, ventilation units, hardware, steering shaft and joints, front seat cushions, suspension bushes, hose clamps, alternators, clutch release bearings
Italy	defroster nozzles and grills, glass, hardware, lamps
Japan	W/S washer pumps, cone and roller bearings, alternators, starters
Netherlands	paints, tires, hardware
Norway	tires, muffler flanges
Spain	radiator and heater hoses, air cleaners, wiring harness, batteries, fork clutch releases, mirrors
Sweden	hardware, exhaust down pipes, pressings, hose clamps
Switzerland	speedometer gears, underbody coatings
US	wrench wheel nuts, glass, EGR valves
England, Germany	muffler assembly, pipe assembly, fuel tank filler
England	steering wheel
England, Germany	tube assembly steering column, lock assembly, steering and ignition
England, France	heater assembly
England, Germany	heater blower assembly, heater control quadrant assembly
England, Italy	nozzle windshield defroster
England, Germany	cable assembly speedometer
Germany	cable assembly, battery to starter
England, Germany	turn signal switch assembly, light wiper switch assembly, headlamp assembly, bilux, lamp assembly front turn signal
England, Italy	lamp assembly turn signal side, rear lamp assembly (including fog lamp), rear lamp assembly
England, Germany	weatherstrip door opening, main wire assembly, tires, battery, windshield glass, back window glass, door window glass, constant velocity joints

(**Table 11.2** continued)

France, Germany	transmission cases, clutch cases
England, Germany	rear wheel spindles
Germany	front wheel knuckle
England, Germany	front disc
England, France, Italy	cylinder head
England, Germany	distributor
US	hydraulic tappet
England, Germany	rocker arm
England	oil pump
Germany	pistons
England	intake manifold
England, Germany	clutch
Germany	cylinder head gasket
England, Germany, Sweden	cylinder bolt
N. Ireland, Italy	carburetors
England	flywheel ring gear

Steel (body steel and forging barstock) from UK, Germany, Belgium, France, Italy, Austria (sheet) and Finland (bar).
Source: Fieleke (1981 : 43).

1930s. The Big Three companies (General Motors, Ford, and Chrysler) lost a combined $3.5 billion, 250,000 workers were given indefinite layoffs, and 450,000 more lost their jobs in industries supplying the major companies (Hill 1984). This crisis has major multiplier effects because the automobile looms large in the US economy. As of 1981, more than 4 million jobs directly depended on it, including 800,000 in automobile manufacturing, nearly 1.5 million in supplier industries, and more than 2.5 million in sales and servicing. Auto manufacturing consumed 60 percent of the country's synthetic rubber production, 30 percent of its metal castings, 20 percent of its steel, and 11 percent of its primary aluminum (US Department of Transportation 1981).

However, the cold winds of global competition have brought a big chill to the US industry. In 1980 imports of automotive parts, engines, and vehicles exceeded exports by more than $11 billion. On top of an overall decline in its world-wide competitive position, the US industry recorded sharp drops in output and employment throughout the period 1977–82 (Trachte & Ross 1985).

The industry's recent contraction is by no means attributable to foreign competition alone. American companies have themselves shifted major elements of production abroad (foreign sourcing) and failed to adapt to higher gasoline prices by building smaller, more fuel-efficient cars. As a result of growing international specialization, international trade in automobiles and automotive parts has grown rapidly. The growth in this trade and the surge in demand for more fuel-efficient cars have given rise to the concept of the "world car" designed to be sold widely throughout the world. Such a car would be produced as well as consumed globally. One example of a car that is assembled from parts made in many nations is Ford's European Escort. Assembly takes place in Britain, West Germany and Portugal from components made in 17 different countries (Table 11.2).

Detroit is a victim of the world car. Its problems therefore are not cyclical or temporary, and not likely to be resolved through a general upturn in the US economy. The strategies that the US automobile companies have adopted in response to changes in the world market – downsizing, computer-aided manufacturing, foreign sourcing, and overseas production – are exacerbating rather than relieving the economic malaise of Detroit (Trachte & Ross 1985).

The city is increasingly isolated from its surrounding area. It has suffered relatively more in terms of available jobs from the decline of the US auto industry than its suburbs or other places where employment in vehicle production was high (Table 11.3). The quality of life correlates with the general malaise. For example, Bunge has dramatically illustrated the asymmetric distribution of life chances within Detroit with his map of infant mortality in the city (Bunge & Bordessa 1975). There is a fourfold range in infant death rates between the city and its suburbs. Central city rates are

Table 11.3 Employment in motor vehicle production: US, Michigan, and Detroit, 1949–82 (thousands).

Year	Total US	Michigan	Detroit
1949	751.3	na	313
1950	816.2	na	334
1951	833.3	na	339
1952	777.5	na	315
1953	917.3	na	365
1954	765.7	na	288
1955	891.2	na	313
1956	792.5	411.7	252
1957	769.3	387.7	252
1958	606.5	288.4	183
1959	692.3	303.4	194
1960	724.1	311.2	198
1961	632.3	268.4	168
1962	691.7	299.2	177
1963	741.3	315.8	192
1964	752.7	332.3	208
1965	842.7	364.5	227
1966	861.6	381.9	236
1967	815.8	361.8	219
1968	873.7	382.8	233
1969	911.4	398.7	246
1970	799.0	333.6	210
1971	848.5	351.8	216
1972	874.8	367.7	224
1973	976.5	399.0	248
1974	907.7	364.3	232
1975	792.4	327.9	199
1976	881.0	352.8	221
1977	947.3	386.5	239
1978	1,004.9	409.6	252
1979	990.4	392.7	232
1980	788.8	333.5	192
1981	783.9	319.4	178
1982	704.8	287.0	167

Note: Data refer to ISIC 371: Motor vehicles and equipment; na = not available.
Source: Trachte & Ross (1985 : 210).

comparable with those of underdeveloped countries such as Peru or Guyana. In contrast suburban rates are comparable to the lowest in the world. The residents of the central city breathe air that is four times as dust-concentrated and contains four times the sulphur dioxide of air in the suburbs. The central city contains much less recreation space, poorer schools, and much higher crime rates (Jacoby 1972, Rose & Deskins 1980).

Today, a majority of Detroit's population is black – about 63 percent in 1980. They now dominate in most areas of the city. Only on the edges of the city where Detroit meets the older suburbs are they less strongly represented or absent. Most of these people, or their parents or grandparents, arrived during and after World War II attracted by the lure of work – and high wages – in the auto industry (Table 11.1). As in other cities in the United States that have large black populations, the predominantly white neighborhoods of Detroit, after becoming open to black residency, have tended to become all or predominantly black rather than remaining "integrated" (Wolf & Lebeaux 1969). The whites have moved out into the suburbs where they are "protected" by higher housing prices and a variety of legal and illegal exclusionary devices (Danielson 1976). Even when blacks suburbanize, they tend to be segregated in little ghettos (Farley 1978).

Rather than the old ethnic attachments to neighborhood, contemporary concerns are much more utilitarian. Wolf and Lebeaux's (1969) studies suggest that concern about the racial composition of schools and appropriate "social behavior" have been at the root of white resistance to integrated residential neighborhoods in Detroit. With attachment to neighborhood now based upon practical individual or family interests, when these interests are threatened people will search for more rewarding residential areas. These are now more often outside the city; to which the city's loss of population testifies.

In the period 1965–7, a series of urban riots rocked the United States. In city after city, black ghettos were racked with looting, burning, sniping, and rioting. Three of the more than 100 riots were major insurrections – Watts, Los Angeles in August 1965, and Newark and Detroit in 1967. In the week of 23–28 July 1967, Detroit was the scene of one of the bloodiest racial uprisings of the 20th century. The week of rioting left 41 known dead, more than 1,000 injured, and 35,000 arrested. Whole sections of the city were reduced to charred ruins. The riot began when police raided a black "after-hours" club and arrested the owner and a number of patrons.

Detroit has in a sense never recovered from the 1967 riot. The city remains physically and politically scarred by it. All sorts of government programs were enacted as an almost direct response to it and other riots. But most of these programs have long since disappeared. The pace of plant closures also picked up (Mollenkopf 1983: 251). Many of the apparent "causes" of the riot remain: high unemployment, discrimination in housing and employment, "hopelessness," and "powerlessness." One of the major changes has been the emergence of a black "power structure" within the city and the election of many black city officials and congressmen. But this group has been constrained to work within parameters laid down by businessmen anxious to avoid such "contentious" places as Detroit.

In state and national politics, Detroit is a major stronghold of the Democratic party. Across elections and offices Detroit as a whole returns Democratic candidates by large majorities (Groop et al. 1984). In the 1980 presidential election, Jimmy Carter carried virtually all the precincts in the city except some nonblack, middle-class precincts on the western and northeastern fringes. The final tally was 78.5 percent for Carter to 17.8 percent for Ronald Reagan. Given low turnouts in the poorer black neighborhoods, this figure would indicate, when compared to the citywide percent black, that the Democratic party still retains an attraction for many of the city's white voters, even after the riot-torn 1960s and the growth of a black political "machine."

Even when a "liberal" Republican governor of Michigan was seeking re-election in 1978 with the "tacit" approval of Detroit's black mayor, his Democratic opponent still won by a margin of 60.6 to 39.4 percent (Groop et al. 1984 : 36). As of 1984, all of the US congressional delegation and all of the state house and senate representatives from Detroit were Democrats. Low turnouts and the persistence of serious economic underdevelopment, however, signal a political malaise in Detroit that will not be readily overcome.

The history of Detroit captures in one place the history of the American automobile industry and the consequences of dependence on one industry. From 1870 until 1910 Detroit was one of several multipurpose cities of the industrial belt. From 1910 until the late 1960s, the automobile industry and war production made Detroit a major national industrial center (Zunz 1982). But the country's sixth largest city has now become enmeshed in what Hill (1984 : 315) terms "a web of uneven development spun first by the flow of industrial and commercial capital to the suburbs and then to the Sunbelt and, more recently, by the reorganization and decentralization of the auto industry on a global scale." This has led to an economically declining and increasingly black city dependent upon large infusions of government funds.

Detroit is now America's equivalent to Glasgow, sharing with that city a precipitous decline in its major industry and the growth of a population with limited job prospects and increasing reliance on government "largesse" and the welfare state. In 100 years Detroit has run the gamut from (a) a multipurpose industrial city with a patchwork of ethnic neighborhoods in which politics was organized around ethnoreligious cleavages through (b), the early automobile era, when occupational divisions displaced ethnic ones in the residential organization of the city and "nonpartisan" (Republican) administration replaced partisan politics; to (c) the period since the 1930s when unionization and the Democratic party became significant but have been followed by a process of suburbanization and a "downsizing" of a much reorganized automobile industry that have together destroyed much of the city's economic base.

Mansfield, Ohio

Mansfield is a small city of fifty-four thousand people in north central Ohio. It is heavily industrial with a large automotive component plant (Fisher Body), a steel mill (Empire-Detroit), and a variety of engineering and metalworking concerns. A number of large industrial employers have closed down over the recent past, notably Mansfield Tire and Rubber and Co. in 1978. Many previously locally owned firms have sold out to conglomerates (Lukas 1984:29).

But the city is located in an agricultural region that has been historically conservative and Republican. In Mansfield itself, registered Democrats narrowly outnumber Republicans, though both are virtually matched by "Independents."

Early in the century the Mansfield region was largely Republican as a result of "sectionalism" within Ohio. Flinn (1962) has pointed to the diverse origins of the original settlers as an explanation. Southern Ohio was settled from the South, and the settlers there brought their Democratic commitments with them. In northern Ohio the settlers, coming from New England, Upstate New York, and Pennsylvania, brought a different set of practices and ideas that became expressed through the Republican party.

This older system only broke down under the impact of industrialization and the emergence of class–ethnic alignments that industrialization and immigration (of Catholic East Europeans, in the main) brought in their train. Burnham (1982:39) argues, however, that "no new political order of similar coherence or partisan stability has yet emerged to take its place." The level of incoherence and instability is indicated by low turnouts in all elections, high drop-offs from presidential to off-year elections, high roll-offs from "high prestige" to "low prestige" offices when voting, and high rates of split-ticket voting. Unlike states like Michigan (including Detroit) and Pennsylvania, but like New York, the New Deal of the 1930s did not bring a political realignment to Ohio: the dealignment from party politics begun in the early years of the century has continued unabated.

In Mansfield the *lack* of strong party affiliations and an active political life are apparent, especially among the working-class (Lukas 1984). There is no mistaking the fact that party lines do closely follow class lines – Lukas could only find three Democrats among Mansfield's 51 most prominent businessmen and professionals – and that both class and party lines follow geographic lines. "By and large, Mansfielders who live south of Park Avenue are middle class and Republican; those who live north of Park Avenue are working class and, at least nominally, Democrats" (p. 30). But many of these Democrats are truly nominal ones. In the past half century, for example, Mansfield has gone for the Democratic presidential candidate on only 5 out of 14 occasions. Republicans have won the majority of the vote on nine occasions.

Yet, for many years Mansfield has been known in Ohio as a militant union town. There have been numerous and bitter strikes over the past fifty years. Lukas (1984) recounts a number that figure prominently in local folklore. In recent years, however, union leaders and union members have become increasingly estranged from one another because, Lukas argues, union

leaders have given in to company demands for wage and benefit concess-
ions. Today, "The tides of change are running against the unions" (Lukas
1984:36).

Blacks number about 16 percent of Mansfield's population but they
account for only about 8 percent of the votes. Since the New Deal those
who voted have voted Democratic, but for many years this was based on a
"brokered" relationship. For many years the local Democratic party dealt
with the city's black population through Mose Haynes, "a black kingpin of
prostitution, gambling and other illegal activities" (Lukas 1984:34). Certainly, the local Democrats have never really "cultivated" the black vote or
appealed to its interests in areas such as discrimination, housing, and
employment. And blacks have been too few to exercise much influence
within the local Democratic party.

The upshot of this divided and politically inchoate Democratic constituency has been a low level of participation, with participation itself based on
a narrow view of individual self-interest rather than any wider ideological
frame of reference. Lukas (1984:29) notes

> Only rarely during three visits did I hear anyone define his position by
> citing the needs of others, or of the nation at large. It is difficult to
> know whether this turning inward is somehow a regional character-
> istic or whether it reflects the prevailing insularity of mid-1980s
> America.

And this was in the context of a serious economic downturn between 1981
and 1983. Unemployment reached a peak of 18.4 percent in February 1983
and is still over 10 percent. Many of the smaller firms are still laying off
workers.

One of the major causes of the lack of mobilization to pursue solutions to
collective problems such as those of unemployment and vulnerability to
plant closings is suggested in the statements of Lukas's (1984:75) "union"
informant: it is the division between workplace and communal politics.
Militancy in the workplace has not been conjoined to a similar militancy in
"politics." Issues in the community relating to housing or social issues
divide those who may have common interests at work. For example, the
New Right has not been inactive in Mansfield. The social agenda of the
fundamentalist and often unaffiliated churches – abortion, homosexuals'
and women's rights, and so forth – has played an important part in dividing
the population (Lukas 1984:44).

The major feature of Mansfield's political life has been its transformation
from a coherent and partisan past to an inchoate and depoliticized present.
Perhaps the increasingly "isolated, atomized, social milieu" of a small city
in an agricultural region with political interests and sentiments antithetical
to those of the city has played a role (Fenton 1962). Lukas (1984:45) quotes
from Louis Bromfield, who wrote of his childhood in Mansfield:

In all the Town and the County and the State, there was nothing of such palpitating interest as politics ... Every small boy bedecked his coat with rows of celluloid buttons bearing the images of his party's candidates from the President to the County Sheriff. There were fist fights and black eyes and on election night the children were allowed to stay up until midnight to hear the first returns. In that Middle Western Country one breathed politics.

But today Esther Hart, a widow aged 59, is more typical. She has never voted. She adds, "My husband never voted either. We were just a family. I don't know, we didn't want to discuss those things. We didn't argue and fuss about it. What's going to be is going to be, whether we vote or not" (Lukas 1984: 45).

Middlesboro, Kentucky and Claiborne County, Tennessee

In the late 1880s, in the vicinity of Cumberland Gap, the famous passageway through the Appalachian Mountains, a phenomenal "boom" created the new industrial and mining center of Middlesboro, Kentucky. In the early part of the century the new roads and railroads of the expanding American economy had bypassed this region because of the difficult terrain. At the time of the Civil War there were only sixty families in the Yellow Creek Valley of Bell County, Kentucky, where Middlesboro was later built, and scattered settlement to the south in the narrow river valleys and mountain valleys of Claiborne County, Tennessee (Gaventa 1980: 49–50). The economy was entirely one of subsistence agriculture, a pattern which in Central Appalachia, according to Owsley (1949: 56–7), creates social isolationism and tight kinship networks.

Most of the highlands of Kentucky and Tennessee opposed secession in 1860, favored the Union, provided manpower to the Union armies, and after the Civil War ended, like the rest of "the South," voted as they had fought – in this case, of course, Republican. The Republican share of the vote in presidential elections has remained high to this day, even through the "realigning elections" of the New Deal (Phillips 1969: 257). The absence of the slavery issue, there were few or no slaves in the highlands, the sentimental partisanship of supporting the Union in the Civil War, and continuing hostility to lowland Democrats in their state legislatures are usually put forward as *sufficient* explanation for the continuing affiliation of the highlanders of Kentucky and Tennessee to the Republican party in national elections (e.g., Phillips 1969).

But in fact the region, and the Republican party, have changed very much since 1865, whatever the merits of the "tradition" argument in terms of the immediate aftermath of the Civil War. In particular, the "isolation" of today is quite different from that of the 1860s. Since the 1880s, agents from "outside" have transformed Central Appalachia. In fact it is from the late 19th century that the contemporary economic and political life of places such as Middlesboro, Kentucky and adjacent Claiborne County, Tennessee can be traced.

Two trends came together in the 1880s to transform the basis of life in and around the Cumberland Gap. One was the increased investment of British capitalists overseas. It was a British company, the "American Association, Ltd.", that built Middlesboro(ugh) Kentucky and established a new industrial order based on coal mining in the surrounding area. A second trend was the increased demand for coal in the already industrialized centers of the United States. In the late 1880s the area around Cumberland Gap, previously held to be unattractive for development, became a "hot property." The catalyst was a Scottish businessman and visionary, Alexander Arthur, who organized the American Association, acquired land and mineral rights, and in only two years, 1887–9, built the town of Middlesboro (Gaventa 1980: 50–8, Caudill 1983: 16–35).

Along with outside control over resources and a rapid penetration of local politics by officials subservient to the area's new "owners" came a new set of values. Most of these related to the glorification by town fathers and mine owners of the ethics of work and conspicuous consumption, and the concomitant degradation of the local culture (Gaventa 1980: 61–8). "Labor stability" was a major concern of the mine owners, and this contributed to the degree of social control they exercised over life in the company towns they established in the vicinity of their mines. Mine owners and managers constantly bemoaned the irregular work habits and high turnover rate of the mining population (Eller 1982). They thus made every effort to maintain a permanent, family-based mining force, and to stimulate "a spirit of contentment with the place." Schools, clubs, and churches were important means of attracting a workforce and creating a degree of stability as well.

In both Middlesboro and in the towns established along with the mines the power of the new "elites" was pervasive. In the mining settlements of Claiborne County, for example, the mine operator was the sole arbiter of justice, regulated access to the town, strictly limited personal and social liberty, and controlled the hub of life in the town: the company store (Gaventa 1980, Eller 1982: 161–98).

This was the setting in which the industrial age came to the Cumberland Gap. Although mountain society had until this time been based upon small family farms scattered along river valleys and in mountain hollows, and there had been few concentrated settlements, the experience of the company town did little to change the lack of communal organization that inhered in the older pattern of settlement. As Eller (1982: 195) puts it,

The mountaineer's primary responsibility had been to himself and his family, and his relationship to neighbors had usually been informal. His experience in the company town did little to change these traditional values, since miners were highly mobile and had no direct political control of their communities through town elections. Life on the farm, moreover, had taught him that his future depended not so much on his own activities as upon the impersonal forces of nature. In the company town, he realized that these impersonal forces lay outside the community – in the decisions of managers in the head office, government policies, and the fluctuations of the coal market. Except for his decision to stay or leave, persons other than himself made the decisions affecting his life. Thus he was individualistic, fatalistic, and present-oriented, and his powerless situation in the company town augmented these traits.

The new values of work and consumption, therefore, melded easily into a matrix of values already in operation. Political quiescence or apparent consensus to the new order was the result. In Middlesboro and Claiborne County in the early years, therefore, political life was completely dominated by the outside interests and the new local elites (Gaventa 1980:75–9).

However, when "the field of power" shifted, as it did in 1892–3 with the temporary bankruptcy of the American Association, Ltd. and the downturn in demand for coal, many miners and the local middle class began to voice challenges. But the "role of powerlessness," as Gaventa (1980:82) puts it, was so strongly internalized that the challenges never amounted to much. Once the company was re-established the opposition evaporated: business returned to what had become normal in the years after 1887.

Middlesboro and Claiborne County survived as industrial outposts despite the "bust" of 1893. But at least until 1929, political quiescence remained a dominant feature of these places. Elsewhere, the picture was often different. When miners were resident in independent communities, as in one place in Claiborne County, and elsewhere in central Appalachia, they were much less subservient (Gaventa 1980:93–5). In the company towns not only the workplace but the other institutions in which collective values could be shaped were encouraged and controlled by the mine operators. The schools were run by the companies who built them, financed by "docking" from miners' wages, and the teachers were hired (and fired) as the local "boss" willed. The churches did serve as an outlet for community expression and as an agent of socialization. But the content of the fundamentalist Protestantism preached within them encouraged servility, fatalism, and individualism (Gaventa 1980:91–2; Eller 1982:161–98).

In the social sphere, and despite the relatively low wages, a number of factors, including individualism and a short-run time horizon, encouraged a great desire for consumption. Perhaps it was only rational to "want" that which "glittered" while one could (Gaventa 1980:91).

In his role as "citizen" as opposed to consumer, a similar phenomenon is visible. Miners had little reason to value the vote. In company towns voting was not a private act. Rather, as the political stakes in electoral outcomes were high for the mine owners, particularly in the elections for sheriff and tax commissioner, sanctions against "deviant" voters were severe. For the miners, encouragement to vote in the "correct" way came in the form of money payments which were made by mine owners upon verification by election officials! (Gaventa 1980: 91).

After 1930 this situation began to change. There was a coal boom in the 1920s, but with the onset of the national depression the demand for coal dropped dramatically. In the face of layoffs and cutbacks the control of the mine owners inexorably waned. A mood of rebellion prevailed. In the spring of 1931, a massive union drive netted many new members in an area previously ununionized; but the resources of the United Mine Workers of America (UMWA) proved insufficient and the leadership too timid in the face of immense repression to maintain a long-term strike (Gaventa 1980: 96–121). More importantly, however, local elites, through their control over sources of information and repression, were able to "isolate" the miners and contain the rebellion. One of the major elements involved in this was the creation and exploitation of the myth that the new militancy was not of the miners' own making or in their interests. It was an "evil" brought in by outside agitators and communists (of whom there were not a few involved in labor organizing). This "demanded the response of militant loyalty to one's nation and culture" (Gaventa 1980: 110). The symbol of the "communist" was especially powerful in undermining miner solidarity. Its meaning was usually interpreted in terms of the local culture, in particular family and church. The miners discovered that the Communist party did not believe in God. Thereafter the term "communist" could always be invoked as another resource in the "mobilization of bias" against social change in Middlesboro and Claiborne County (Gaventa 1980: 104–21).

Since the 1930s the dual pattern of "normal" quiescence and periodic challenge has by and large continued. The years during and immediately after World War II were ones of relative prosperity. In 1950 in the Clear Fork Valley section of Claiborne County, the main mining area, there were ten underground mines, producing just under a million tons of coal. These mines employed a total of 1,400 men, with no mine employing fewer than 50. The mining communities contained some 30,000 people with some 10,000 in the central town of Clairfield. Middlesboro, the major trading center for the Cumberland Gap area, had a population of about 20,000, more or less the same as in 1894 and today (Gaventa 1980: 125).

The relative prosperity did not last long. Beginning in 1954 the mines began to close. By 1957 there were only 230 mining jobs in Claiborne County, and by 1960 only 144, compared to 1,230 in 1952. Entire towns were emptied of people as their residents migrated to the cities in search of work, part of the one million migrants from central Appalachia during the 1950s.

For those who remained, the type of mining changed from the underground mines to "dog hole" mines, as people tried to carve a living from hillsides, or strip mines which involved bulldozing the mountainsides by machine rather than by human labor. By 1969 Clairfield had only 2,000 people, and the population of Clear Fork Valley had declined from 40,000 to 12,000. Fully 95 percent of the families in the valley lived below the official "poverty line." Beginning in 1970, coal production began to increase as demand increased. Today production is much higher than it was in 1950, but with a much reduced number of miners (Gaventa 1980: 126–7).

In the 1960s, outside agencies, such as the Appalachian Regional Commission established in the wake of John F. Kennedy's encounter with poverty during the 1960 West Virginia presidential primary, argued that the problems of places such as Claiborne County were caused by "isolation." The solution was to build highways to "integrate" these places into the national economy. But the Appalachian coal valley, such as Clear Fork Valley, is not and has not been isolated since 1880. It is an important part of the national and international economy. The owner of the land and the coal, the American Association, Ltd., of London, is part of a large financial conglomerate. The coal is mined by Consolidation Coal Company, a wholly owned subsidiary of Continental Oil Company. The coal taken out of the valley goes to Georgia, Mississippi, and Alabama, where it is burned to produce electricity for the urban, industrial centers of the South.

The prevalence of political quiescence has persisted along with the coal industry. Gaventa (1980: 141–64) shows that the *potential* issues of the Clear Fork Valley – undertaxation of the landowner and coal company, the environmental degradation caused by strip mining, job vulnerability, low standard of living – are nonissues in the local political process. On election day the people dress in their "Sunday best" and go to the schoolhouse to vote. Local contests receive most attention. Party identification is rarely mentioned. Local "factions" headed by competing elites appear to be the basis for political cleavages. In presidential and congressional elections the Clear Fork Valley votes Democratic, as it has since the bitter days of 1932 and the coming of the New Deal. In this respect it differs from surrounding farm communities and local trading centers such as Middlesboro where the traditional mountaineer ethos and middle-class moralism, respectively, give the Republican party the edge. But national politics appears to strike little interest. And in local politics "public challenges by the non-elite, as candidates or critics, are deterred by feelings of inadequacy, fear of reprisal, or simply the sense that the outcome of challenge is a foregone conclusion" (Gaventa 1980: 144). The issues are those the elites choose – personality, charges of corruption, "dirty tricks," and so forth.

Table 11.4   Voting for "company" candidates, by community (percent).

	Company town	Coal community	County seat	Rural community
county judge (1970)	100	79	42	65
school board (1970)	94	31	63	28
sheriff (1968)	97	61	51	39
schoolboard (1966)	88	40	57	40
tax assessor (1940)	89	73	54	52
sheriff (1940)	94	70	46	46

All voting data are from original Claiborne County Tally Sheets, as filed with the Tennessee Secretary of State and recorded in the State Archives, Nashville, Tennessee.
Source: Gaventa (1980 : 149).

Gaventa (1980 : 145–64) presents evidence from a number of elections in the period 1940–70 to support his "vulnerability thesis": that the people who are most dependent on the companies are the least independent voters. He demonstrates that people living in a "company town" are less independent than those in a "coal community" that is not totally company-dominated, who in turn are less independent than those in the "county seat" and a "rural community" where company influence is much less important (Table 11.4). This is irrespective of party labels. Some "company candidates," such as the County Judge (1970), have been Democrats; others, such as the School Board candidate (1970), have been Republicans. Though the latter's "Republican" designation does seem to have undermined his performance in the coal community, it did not help him in the typically more Republican rural community.

Much of the literature on single-industry communities, especially mining areas, suggests that their social homogeneity, close interrelationship between workplace and residence, and the regular – and dangerous – routine of work inevitably lead to a social solidarity and political radicalism not found in other places. But this emphasis misses the importance of what Gaventa (1980) calls the "power field" around the occupational community. In certain settings, the homogeneity of mining communities *can* lead to enhanced solidarity and radicalism – this even happened on occasion in Middlesboro–Claiborne County – but often homogeneity can mean enhanced vulnerability. This is particularly apparent with respect to local elections where most is at stake for local elites. Even turnouts for local elections can be higher in the more "vulnerable" settings (Gaventa 1980 : 150–61).

The long-established tradition of political quiescence in Middlesboro–Claiborne County shows few signs of abeyance. To the contrary, and despite the national social reforms of the New Deal (late 1930s) and the War on Poverty (late 1960s), the nature of political life in this part of Central

Appalachia continues much the same as it has ever since 1887, with the exception of those "bursts" of resistance in the early 1890s and the early 1930s. In other parts of the region this pattern has not always been repeated. For example, as Gaventa (1980) points out, miners in other places have been more militant and for longer periods of time. He argues that this depended critically on a lesser degree of company control and paternalism than that exhibited in Middlesboro–Claiborne County.

The potential power of the miners in Middlesboro–Claiborne County, therefore, has been crucially constrained by the removal of the miners' issues from the local political agenda, and the ways in which local elites have been able to shape the miners' conceptions of what is *possible* for them to achieve. The *possibilities* of power, however limited, have been effectively denied to people whose backgrounds and historical experience have not prepared them to exercise it. They remain dependent, prisoners of a social structure which through their quiescence they continue to reproduce.

Miami, Florida

With the exception of a few mountain cities (such as Knoxville), most southern urban centers were strongly Democratic in the period between the end of Reconstruction and the First World War. They followed much the same politics as their rural hinterlands. But some of the southern cities best known today, especially in Florida and Texas (for example, Houston, Dallas, and Miami) were largely swamp or pastureland in 1914. By 1954, however, these places had all grown phenomenally. In the case of Miami, the boom began in the 1920s. Growth was financed by real estate speculation to meet the demand for retirement homes and the vacation facilities based on southeast Florida's year-round tropical climate. Much of the incoming population came from the North, especially New York and New Jersey, and had nothing in common with the rural population of northern Florida.

Miami grew from a population of 29,571 in 1920, to 110,637 in 1930, 249,276 in 1950, and 291,688 in 1960. Since 1960, much of the growth has been in the suburban areas around Miami: in 1980 the city itself had a population of 400,000 in a metropolitan area (Miami–Dade County) of 1.8 million (Phillips 1969:272, Nordheimer 1985a).

In the years through the 1950s, population growth came largely through the migration of retirees and the settlement of people arriving to work in the businesses serving retirees and vacationers. Miami and the Miami area still have large numbers of senior citizens who are sensitive to issues and policies that especially involve older and retired people.

In the 1920s the newcomers gave large majorities to Republicans in state and national elections. Many of them were "transplanted" northern Republicans. Others moved towards the Republican party because of the perceived differential between themselves and the "local" Democrats. In the 1928 presidential election they helped lay the foundation for Hoover's "upset victory" in Florida over Al Smith (Phillips 1969:271). The Depression limited the Republican vote, however. In 1932 only 34 percent of the Dade County vote was Republican. Florida did not give a majority to a Republican presidential candidate again until 1952.

During and after World War II, the Miami boom regained its momentum. New resort and retirement communities sprang up all along the coast of central and southern Florida, catering to northern middle-class retirees and immigrants. The "new" Miami, like the new cities of Texas and southern California, represented a new urban America. Without heavy industry, and aggressively and monotonously middle class, these new cities were expanding at the same time that America's industrial cities, such as Detroit, were beginning to stagnate. Eisenhower in 1952 and 1956 and Nixon in 1960 cashed in on the votes of Miami's "middle American" majorities. Goldwater lost this support in 1964, largely because of his hostility to the Social Security system. In 1968, although the Republican party made an extraordinary comeback in Florida as a whole, Miami–Dade County remained with the Democrats. Why was this so?

Some part of an explanation rests on the changed composition of Miami's population from the 1950s on. In those years an increasing proportion of the migrants coming to greater Miami were Jews from New York and other northern cities. Miami Beach – the coastal strip to the east of the city of Miami – became one of the most heavily Jewish cities in the United States (Banfield 1965). Political changes and revolutions in Latin America introduced other distinctive groups of migrants. The most important of these were the Cubans who flowed into Miami after the Castro Revolution in Cuba in 1959. Though some of these Cubans were far-right conservatives, most at first blended into the existing Hispanic population of Miami. Many did not vote, however (Banfield 1965:104). These demographic changes served, in Phillips's (1969:284) words, "to make Miami into an atypical – and atypically liberal – Southern city."

In 1957, Miami–Dade County became the first metropolitan area in the US to acquire a "metro government." By only a bare majority of the 26 percent of Dade County's registered voters who voted, a Dade County home rule charter was approved by popular referendum on 21 May 1957 (Sofen 1963). County government became in effect a metro government. This government assumed a number of functions performed by the 26 local governments in the metropolitan area including sewage, water supply, transportation, traffic, central planning, and "those municipal functions susceptible to areawide control." This vagueness over precisely what is an areawide problem has led to a great deal of litigation over the powers of the metro government. The Miami metro plan included council-manager government for the county. This was championed by Miami's business

leaders. But many other groups have been opposed to both metro government and the council-manager form it has taken. For example, the cities of
Miami Beach and Hialeah have been strongly opposed to the metro idea.
This is also true of municipal employees, labor unions, local chambers of
commerce, and the local press (except *The Miami Herald*). A number of
referenda designed to amend the charter in order to cripple metro government have been defeated by only slim margins.

Sofen (1961 : 20) has explained the narrow victory of the "metro" proposal
in Miami in a way that highlights the nature of politics there in the late 1950s
and early 1960s:

> Greater Miami is unique in a number of ways; it is warm when almost
> every other place in the nation is cold – a fact which has attracted a
> constant influx of migrants from all over the nation . . . Miami is almost
> wholly devoid of strongly organized political factions and strongly
> organized labor or minority groups. It is generally lacking in organized
> and consistent community leadership . . . Miami with its "every man
> for himself" type of politics in effect has a no party system and,
> consequently, was spared the kind of struggle that might have occurred
> if the fate of political parties had turned on the outcome of the move to
> create a metropolitan government.

Since the late 1960s, Miami has undergone something of a metamorpho-
sis. Change has come in two major ways. First of all, demographic
transformation has continued at a rapid pace. Today, 650,000 Cuban-
Americans live in a Dade County of 1.8 million people. Over 50 percent of
Miami's population is Cuban, and Cubans constitute 40 percent of the city's
registered voters. Most of the rest of the population is evenly divided
between whites and blacks, with a small percentage of Puerto Ricans and
other people of Hispanic origin (Nordheimer 1985a). Of the Cuban population, 125,000 arrived in the famous "Mariel" expulsion of 1980.

Until recently Cuban votes did not lead to Cubans being elected to office
even though local politics, as described by Sofen, was "wide open." Though
Cubans have been highly successful in banking, commerce, and law, they
made slow headway on the political front. Only in 1985 was a Cuban elected
mayor of Miami for the first time. The election of Xavier Suarez symbolizes
both the coming to power of Cubans in Miami and their lack of need for
support from other groups: in a runoff election against another Cuban-
American, Suarez relied heavily on Cuban voters who turned out in
sufficient numbers to guarantee victory. His opponent garnered most of the
black vote, in a low turnout (Nordheimer 1985b).

Mayor Suarez also symbolizes an increasingly dominant political outlook
on the part of Cuban-Americans in Miami. Americanized to a degree, they
are still steeped in Cuban culture and the Spanish language, fiercely
anti-communist, and tending to align themselves with the Reagan wing of

the Republican party (Nordheimer 1985b). There is consequently an increasing divide between the Cubans and other groups in the city who tend to be consistently more liberal on a wide range of issues. These groups, for example, backed the maverick "liberal" candidacy of John Anderson in the 1980 presidential election to the extent that he garnered over 10 percent of the vote in Miami–Dade County. There is particularly heavy resentment on the part of many "Anglos" (whites) and blacks over what they see as the "Latinization" of Miami. This peaked in 1980 when Dade County voters passed an ordinance forbidding the use of any language other than English in the conduct of official business.

Phillips (1969), writing in the late 1960s, missed the essential differences between the Cubans of Miami and other Hispanics. The Cubans came well equipped to seize economic opportunities in the United States; because of the nature of the Cuban Revolution it was the educated, professional, and managerial classes who filled the ranks of those who fled from Cuba (Nordheimer 1985c). This aggressive middle class was temperamentally suited and educationally prepared to adjust to the situation in Miami. Given the circumstances under which they left Cuba, they have become willing converts to the "new" Republican party and Ronald Reagan's view of the world – as witnessed to by the "Cuban vote" in the 1980 and 1984 presidential elections.

A second change has been the emergence of Miami since the late 1960s as the "gateway" to Latin America from the United States. This has had both positive and negative features. When the narcotics boom took off in the mid-1970s, Miami became the drug capital of the world. More than 70 percent of the US supply of heroin, cocaine, and other illegal substances flows through it. This traffic has brought drug-related crime (Miami's murder rate is the highest in the country), and wealth. The influx of "hot dollars" has made Miami a rival to New York City as the nation's financial capital (Lernoux 1984).

Miami boomed throughout the late seventies. Many banks and multinationals were lured to Miami by its huge cash flow. Not all of this came directly from drug smuggling and money laundering: in particular, exporters, retailers, and realtors catered to the more than two million Latin American refugees and tourists who flocked to the city each year from 1976 through 1983. With its over 50 percent Hispanic, predominantly Cuban, population, Miami became Latin America's playground and political refuge. Argentinians and Venezuelans bought many luxury condominiums, gambling that the city's real estate market was a better investment than their own inflating currencies. For those on extended shopping trips, there were cameras, watches, televisions, and designer blue jeans. If these could not be carried home they could be shipped through the convenient Port of Miami container terminal. Exports through Miami reached almost $8 billion in 1981, about 80 percent going to Latin America. For those interested in "dollar salting" Miami became a substitute for Zurich. During the boom in lending by American banks to "Latin America" in the late 1970s, many American banks set up branches in Miami to capture their share of the $3 billion to $4 billion fleeing Latin America each year (*Business Week* 1984).

These infusions, legal and illegal, transformed Miami. From a city historically dependent on the pension incomes of its large population of retirees and the spending of tourists from the North flocking to sun and beaches, Miami became in ten years a Latin American Hong Kong. The prime customers became Latin tourists and Latin business.

The boom changed the orientation and the skyline of Miami as well. The center of activity moved away from the hotels on the beach to a downtown exploding with banks and other high-rise buildings. The most prestigious banking address is Brickell Avenue with its gold-and-black skyscrapers. Coral Gables, where a hundred multinationals have their Latin America headquarters, has also become a chic location. When added to the more than 100 banks representing 25 countries, the scale of Miami's financial activities comes into focus (Lernoux 1984).

In 1982 the Latin tourist boom collapsed. The world recession of the early 1980s, the ending of big US bank loans, and the austerity packages imposed by the International Monetary Fund have led to a devastation of the Latin economies. Exports through Miami plummeted by 27 percent ($2 billion) from 1982 to 1983. The great dream of Miami as a gateway to Latin America faded. But the banks have not moved. Only a few multinationals have left town. There has been no mass exodus. The companies that have left do not blame their departure on the Latin American collapse. General Electric, for example, decentralized its Miami operation to separate country headquarters in Mexico, Venezuela, and Brazil (*Business Week* 1984). Banks and business in general are awaiting a revival of the Caribbean and South American economies.

In the meantime there are still the transfer payments – pension checks, Social Security, and retiree saving accounts and, of course, the drug money. It will not be easy for the city to go back to dependence on retirees and American tourists. In particular, Miami's vacation and retirement image has been repeatedly tarnished by crime and interethnic conflict, as exemplified by the riots in the black neighborhood of Liberty City. Despite massive federal enforcement programs and frequent "declarations of war" on the drug traffic, Miami remains the drug-smuggling capital of the US. Too many people, including those in respectable organizations like those banks waiting for Latin American recovery, have too much at stake in the drug business (Lernoux 1984).

Miami, a modern city by any standards, having its roots only in the 1920s, has undergone a remarkable transformation since the late 1960s. From a predominantly retirement and vacation city it has become a major entrepot in the flow of goods and services, legal and illegal, between the United States and Latin America. At the same time Miami has become something of a "Cuba-in-exile" for refugees from "Revolutionary" Cuba. This immigration has sent shockwaves through the social and political life of the city. In 1968, Miami's Cubans may well have appeared to be just another group of Hispanics, but as of 1985 they seem quite distinctive from the Mexican Americans and Puerto Ricans. Miami today is not the transplanted northern "liberal" city of yesterday: neither is it a Mexican-American San Antonio or Puerto Rican Bronx (New York City).

Conclusion

The major thesis of this chapter, as of Chapter 8, is that the "factors" causing political behavior cannot just be added up in linear fashion (census class, census age, census ethnicity, etc.) to constitute an adequate explanation. To the contrary, it is how these factors "come together", take on meaning for people, and determine political outcomes that constitutes a satisfactory political analysis. In other words, it is in places that causes produce the reasons that produce political behavior. Many of these causes, as I have argued, emanate from other places beyond the confines of the locality. But it is in the locality and through the choices of local populations that the various causes of political behavior, from the shifting economic position of the US automobile industry, in the case of Detroit, to Castro's Revolution in Cuba, in the case of Miami, work to structure political expression. This is what "structuration" is all about.

12 *Place and American politics*
Place and political mobilization

The concept of political mobilization is critical in providing a link between places, on the one hand, and the state and organizations such as political parties that provide "bonding" for the state, on the other hand. Modern territorial states are integrated politically to the degree that they can gain legitimacy and extract allegiance from their inhabitants. Political parties are important agents of this task. In particular, they provide a mechanism for representing and "brokering" conflicting place-based interests. In modern politics, acceptance of the legitimacy of the political regime depends in large part on the successful performance of this task. Of course, political parties do fulfill other tasks for the state: they recruit people for political office, they educate or socialize their adherents into a particular interpretation of the state and what it can do for them, and they also, on obtaining office, perform a policy-making function *if* they attempt to put a political program into operation.

In the present context, however, it is the nation-building or integrative task that is of primary significance. But there is always a tension intrinsic to this, as pointed out in Chapter 3. There is a conflict between the popular pressure for state intervention in the form of redistributive and welfare policies, and elite pressures for capital accumulation and economic growth. Legitimacy has a price. Historically, American political parties have been able to rely on a widespread acceptance of the state as an *intruder* into people's lives to minimize their importance as redistributive agents (Kleppner 1982:160–1). This has been possible because, compared to Europe, there has been a dominant *antistate* intellectual–cultural tradition in the United States reproduced within society by such instruments of cultural hegemony as the churches, schools, conscript army, the communications media, other institutions, and the tradition of individualist–liberal capitalism. This tradition has systematically devalued the importance of the state as an agent of redistribution.

There have been two historical conjunctures that have produced this. First, America was in one sense "born free." It had unique advantages in its founding – historical timing, exploitable natural resources, lack of feudal institutions and hence no "struggles over the state" between nobility and bourgeoisie and its founding peoples, many of whom were overtly enthusiastic about the virtues of liberal capitalism. It became, in Friedrich Engels's words, "a capitalist paradise" (Burnham 1984:114). Secondly, the emergence in this century of the United States as a dominant actor within the world political economy produced both an affluence-generating boom between 1945 and 1967 and an intense nationalism based on American ascendency to "world power" status (Agnew 1987a). This reinforced the sense that "America delivers," come what may.

Without the continued operation of these historical resources, however, American parties are faced with the accumulation–legitimacy tradeoff. This has been the case during certain "crisis" periods such as the 1890s, the 1930s, and, especially, since the late 1960s. It is at times like these that political parties become important as "back up" for a jaded and fading hegemony. The problem is that the political parties may themselves be increasingly unacceptable to large numbers of people. For example, nearly half of the American electorate does not now participate in elections. There is a reasonably high probability that these nonvoters are people whose interests cannot achieve organized expression as long as the uncontested cultural hegemony around which the major political parties organize themselves survives. The parties are also increasingly problematic to the "active electorate." Blumenthal (1982) has proposed that parties are in the process of "marginalization" within American politics, and as a consequence there is an increasing disintegration of the national polity. Given the importance of political parties as national integrating mechanisms, a "crisis of the state" may be in the offing (Phillips 1982, Burnham 1985).

In this chapter the mobilization crisis theme is pursued by examining the extent to which the major American political parties are failing to perform their "integrative function": bringing together places into one national framework. This is done by (a) detailing the levels of participation in electoral politics in the United States and showing how these correlate with "crises" of the state and (b) examining the extent to which parties are losing their ability to organize elections as national "integrative" contests. The focus in both respects is on place, participation, parties, and integration.

Place and participation

The rationale for examining electoral participation lies in its "systemic" importance. Voting is a major indicator and contributor to the integration and maintenance of the state and its constitutional arrangements. Participating in national elections is an affirmation of faith in the meaningfulness of the institutions in which the elections are embedded (Nettl 1967). In the United States, voting in national elections is the participatory act most available to the largest number of people. It requires the fewest resources of all modes of political participation, and it represents the most widely accepted definition of citizen influence upon government (Devine 1972:42). For large numbers of people, voting is the only form of political participation in which they engage (Verba & Nie 1972:79). Many people also believe that elections have a "great deal" of impact on public policy (Kleppner 1982:5).

From one point of view, electoral participation should have increased substantially over the course of American history. Despite the rhetoric of the Founding Fathers, they had no commitment to universal suffrage. They offered Americans only a limited property-based franchise. But through the struggles of the early years to extend the franchise to all native white adult males and in later years to blacks, aliens, women, and young adults, voting rights were made more widely available (Williamson 1960, Lawson 1976). With the exception of actions early in this century against aliens and southern blacks, the general thrust of changes in the suffrage since the Civil War has been in expanding the franchise. Residence requirements, literacy tests, and voter registration laws have served to limit electoral participation compared to many other countries. They were designed to exclude or limit participation by the poor and recent immigrants. But over the recent past even these barriers have receded (Kleppner 1982:10). Voter registration may not be the major barrier to electoral participation it once was (Burnham 1982:139–40).

However, electoral participation has *decreased* rather than increased since the 19th century (Burnham 1982, Kleppner 1982). There has been a long-term decline in the level of voter mobilization. To obtain some sense of the magnitude of the decline in electoral participation, in this case in presidential elections, it is important to make a regional analysis of turnouts, expressed as percentages of the potential electorate (Table 12.1). This is because barriers to participation and levels of interparty competition have differed between regions. For example, the southern states excluded blacks and some others from voting through a variety of measures not found elsewhere in the United States from Reconstruction through until the passage of effective voting-rights legislation in 1965. The dominance of one party in the region, the Democratic, throughout this period, also lowered turnouts by eliminating any real political choice. The re-establishment of the Republican party in the South and the enfranchisement of blacks have led to increased turnouts in that region. Continuing *declines* in participation elsewhere in the country have created a "nationalization" of turnout rates, but at much lower levels than in the previous period of nationalization, 1860–76.

But if participation is based on place-specific conditions and decisions, a more "microscopic" analysis is in order. This can be approached by examining presidential and off-year congressional turnouts by state for two periods: 1874–92, when turnouts peaked in the US (except in the South); and 1952–70, when turnouts entered their most recent phase of decline (Table 12.2) (Burnham 1982:132–6). Turning first to the 1874–92 turnouts, a number of features are apparent. First, the highest turnouts are generally in the most industrialized and urbanized states (mainly in the Northeast and near Midwest). Of the fourteen whose turnouts ranged from 80.1 to 92.7 percent

Table 12.1 The appearance and disappearance of the American voter: regional and national turnouts in presidential elections, 1824–1980 (percent).

Year	North and West	Border[a]	Non-South	South[b]	National
1824	25.5	35.0	27.1	27.4	27.2
1828	60.6	68.4	61.6	42.6	57.3
1832	64.8	62.7	64.6	30.1	57.0
1840	82.0	77.4	81.5	75.2	80.2
1844	80.2	79.8	80.1	74.3	78.9
1848	74.5	70.2	74.0	68.2	72.7
1852	73.7	59.9	72.0	59.4	69.4
1856	82.9	66.7	81.0	72.0	79.2
1860	84.6	73.3	83.3	76.3	81.9
1864	80.8	43.2	75.9	–[c]	75.9
1868	86.4	58.5	82.8	71.5	80.9
1872	75.3	66.6	74.0	67.0	72.3
1876[d]	87.0	78.7	85.8	75.3	83.3
1880	87.6	77.4	86.1	65.1	81.0
1884	84.8	76.6	83.7	64.3	79.1
1888	86.3	83.7	86.0	63.8	80.8
1892	81.5	78.5	81.1	59.2	76.1
1896	85.6	89.3	86.1	56.9	79.5
1900	82.3	85.7	82.7	43.4	73.7
1904	76.6	76.7	76.6	29.0	65.6
1908	75.6	79.2	76.1	30.8	65.8
1912	67.1	71.4	67.6	27.8	59.0
1916	68.1	76.6	69.1	31.7	61.7
1920[e]	53.6	63.6	54.9	21.7	49.2
1924	57.4	58.3	57.5	18.9	48.9
1928	66.7	64.5	66.4	23.5	56.9
1932	66.3	65.7	66.2	24.3	56.9
1936	71.8	68.0	71.4	25.0	61.0
1940	73.7	67.4	72.9	26.1	62.4
1944	66.8	56.4	65.6	24.5	56.3
1948	62.8	54.2	61.8	24.9	53.4
1952	72.1	66.3	71.4	38.4	63.8
1956	69.8	64.2	69.2	36.6	61.6
1960	73.6	66.0	72.8	41.4	65.4
1964	69.6	61.5	69.0	46.4	63.3
1968	66.6	59.8	66.1	51.8	62.3
1972	61.7	55.1	61.1	45.7	57.1
1976	58.8	54.5	58.5	48.8	55.8
1980	57.3	54.0	56.9	50.0	55.1

[a] Border states: Kentucky, Maryland, Missouri, Oklahoma, West Virginia.
[b] Southern states: Alabama, Arkansas, Florida, Georgia, Louisiana, Mississippi, North Carolina, South Carolina, Tennessee, Texas, Virginia.
[c] Civil War.
[d] Last Reconstruction Election.
[e] General Woman's Suffrage.
Note: Throughout, aliens are excluded from the population-base denominators on which these estimates are based.
Source: Burnham (1981b : 101).

Table 12.2 Mean turnout in presidential and off-year congressional elections, 1874–92 and 1952–70.

Rank (1874–92)	State	President	1874–92 off-year congressional	Mean drop-off	President	1952–70 off-year congressional	Mean drop-off
1	Ind.	92.7	83.5	9.9	73.5	59.9	18.5
2	NJ	92.2	76.6	16.9	69.6	49.7	28.5
3	Ohio	92.1	76.8	16.6	67.1	48.7	27.4
4	Iowa	91.8	73.4	20.0	72.9	49.0	32.8
5	NH	89.7	83.0	7.5	74.7	53.8	27.4
6	NY	89.0	68.4	23.1	66.0	49.5	25.0
7	W.Va.	87.5	72.6	17.0	74.2	50.5	31.9
8	Ill.	86.1	67.4	21.7	73.4	54.2	26.2
9	NC	84.5	67.3	20.4	51.8	28.8	44.4
10	Conn.	82.8	69.7	15.8	74.8	61.2	18.2
11	Pa.	82.7	70.5	14.8	66.7	54.8	32.8
12	Md.	81.4	65.3	19.8	56.1	37.7	32.8
13	Wis.	81.3	64.4	20.8	69.6	50.1	28.0
14	Kans.	80.1	66.6	16.9	67.6	50.3	25.6
15	Mo.	78.2	64.9	17.0	68.7	42.3	38.4
16	Fla.	76.6	67.9	11.4	49.4	25.1	49.2
17	Va.	76.4	54.2	29.1	37.6	22.5	40.2
18	Ky.	76.3	45.6	40.2	56.8	30.7	46.3
19	Del.	76.1	65.8	13.5	72.7	56.1	22.8
20	Mont.	74.2	70.0	6.0	70.8	61.9	12.6
21	Nev.	73.9	73.4	0.7	62.5	50.9	18.6
22	Cal.	73.4	65.1	11.3	65.9	53.1	19.4
23	Vt.	73.3	56.2	23.3	67.2	53.5	20.4
24	Me.	73.2	70.1	4.2	66.1	49.2	25.6
25	Tenn.	72.7	54.9	24.5	49.1	28.0	43.0
26	Tex.	72.3	54.9	24.1	43.2	20.5	52.5
27	Mass.	71.8	58.2	18.9	71.7	54.7	23.7
28	Mich.	71.5	63.3	11.5	68.7	51.3	25.3
29	SD	70.7	80.4	+13.7 Acc	75.3	61.5	18.3
30	Minn.	70.3	60.3	14.2	72.8	58.5	19.6
31	Wash.	67.3	48.2	28.4	69.4	50.2	27.7
32	Neb.	66.1	55.2	16.5	67.6	50.3	25.6
33	Ore.	64.9	70.2	+8.2 Acc	69.0	55.0	20.3
34	Idaho	61.1	67.0	+6.2 Acc	76.4	63.0	17.5
35	Ala.	62.2	47.0	24.4	34.4	25.4	26.2
36	Ark.	61.4	34.9	43.2	44.3	36.9	16.7
37	SC	58.2	54.8	5.8	34.0	19.9	41.5
38	ND	56.6	66.4	+17.3 Acc	73.8	55.4	24.9
39	Col.	55.4	51.4	7.2	69.7	52.9	24.1
40	La.	54.6	49.7	9.0	44.7	17.6	60.6
41	RI	52.5	26.4	49.7	73.0	59.0	19.2
42	Ga.	48.9	30.6	37.4	37.2	19.5	47.6

Table 12.2 (*Continued*)

Rank (1874–92)	State	1874–92 President	1874–92 off-year congressional	Mean drop-off	1952–70 President	1952–70 off-year congressional	Mean drop-off
43	Miss.	48.3	35.3	26.9	31.6	16.7	47.2
44	Wyo.	47.7	49.1	+2.9 Acc	70.6	60.5	14.3
Admitted since 1892							
	Ariz.				52.6	41.1	21.9
	NM				60.9	50.9	16.4
	Okla.				63.3	40.9	21.2
	Utah				78.4	63.1	19.5
	Alaska				53.4	43.6	18.4
	Hawaii				55.0	47.4	13.8
Subtotals							
North and West		85.4	70.8	17.1	68.5	52.6	23.2
Non-South		84.5	69.3	18.0	68.0	51.2	24.7
Border		79.1	60.3	23.8	63.2	39.7	37.2
South		65.6	49.9	23.9	42.6	23.3	45.3
Total USA		78.5	64.7	17.6	62.0	44.5	28.2

Acc = newly admitted state
Source: Burnham (1982 : 134–5).

(more or less equivalent to recent turnouts in France, Italy, and other European countries), only two are rural states (Iowa and Kansas). Rhode Island is an anomaly in that though it was highly urbanized it retained property qualifications for voting until the 1920s. However, at this time, urbanization was strongly correlated with voter mobilization.

Second, the shift in the South towards low rates of participation was under way. Six out of the eleven ex-Confederate states were in the lowest quartile. But there were exceptions. Florida, Virginia, and North Carolina had high rates of participation in presidential elections.

Third, and finally, the "drop-off rate" between mean presidential and mean off-year congressional turnouts was relatively low, especially outside the South. This is an important indicator of partisan mobilization and commitment to participation across elections.

A very different picture emerges from examining the 1952–70 turnouts – even excluding as this period does the even lower turnouts of the 1970s. First, turnouts in presidential elections in this period are lower than off-year congressional turnouts in the earlier period. Secondly, drop-off rates are higher across the board in every state (except the "special" case of Rhode Island). This is so even though many states now have gubernatorial elections in these off-years, when in the past they did not. One might have expected this to stimulate participation in off-year congressional elections. Thirdly, and finally, there has been a major change in the states in the top and bottom quartiles of participation. At the top the Northeast is represented by only Connecticut and Rhode Island. If Illinois (actually marginal to the top quartile) were included, it would be the only midwestern state. All the rest are found in the sparsely populated and peripheral areas of the Plains, the Rocky Mountains, or the Pacific Northwest (Oregon).

Verba and Nie (1972:242–7) suggest that it is precisely in large cities and most suburban areas that the *lowest* levels of all forms of political participation, including voting, are now to be found. Self-contained, isolated, closely bounded communities are the ones in which rates of political participation are highest. These are predominantly found in the sparsely populated and peripheral parts of the United States. Verba and Nie stress the mutually reinforcing effects of residence and workplace in such places for increasing political participation (p. 243): "Some towns or cities form the center of the lives of their inhabitants – they live and work there, engage in social life there, and their children go to school there."

However, this cannot be all there is to it. Location, as well as locale, is at issue. Other parts of the country characterized by such places have had exceedingly low turnouts. The southern states, Arizona, and Alaska constituted the bottom quartile in 1952–70. The southern "pattern" has remained much as it had been tending in the late 19th century, except that turnouts improved over the extreme lows of 1900–48.

The critical question asked by analysts such as Burnham and Kleppner has been: what happened to produce such a long-term net decline in turnouts and the peculiar geography associated with it? One important cause was the introduction from the 1890s onwards of a whole variety of barriers to electoral participation. Many of these were put in place by locally dominant parties or coalitions and designed to restrict electoral competition. They were a direct response to the electoral challenges to the dominant liberal–capitalist order that emerged in the 1870s and 1880s along with high rates of electoral participation. The most important electoral "reforms" were personal registration, elimination of alien voting, literacy tests, personal periodic registration for cities only, poll taxes, "white primaries," and direct primaries (Burnham 1982:138).

The net effect of the changes was the creation of a very peculiar national political system from 1896 to 1932. This was one in which party competition was suppressed and two party hegemonies emerged: the Democrats in the South, and the Republicans in the North and West. Both rested, in a sense, on the huge mass of nonparticipants. A solution thus emerged to the problem of "winning" a legitimate acquiescence to the capital accumulation goals of two regional elites without a challenge "from below." As Burnham (1982:142) puts it, "a 'modern' American electorate had come into being i.e., one which was heavily class-skewed in participation *in the absence of an organized (or organizable) socialist mass movement which could mobilize lower-class voters in modern industrial-urban conditions*" (emphasis in original).

The only places where an exception appeared to the rule of low levels of working-class participation were cities where "political machines" were well-established and able to withstand the pressures for "reform" against them. In Philadelphia, for example, in elections from 1910 until 1926 wards where the machine was entrenched delivered turnouts 30 percent or more higher or more than statewide averages. These high turnouts declined, however, when machine interests were not involved (Burnham 1982:143).

The decline in turnouts was arrested on a wide scale only when the Great Depression destroyed the ideological and electoral base of the Republican party in the northeastern and midwestern cities. The period 1932–40 was marked by the emergence of the Democrats as the new majority party nationwide, with a parallel mobilization of the electorate. Though mobilization of nonvoters was a major feature of the "New Deal tide," conversion of previously Republican voters was also involved (Burnham 1982:146–7, Kleppner 1982). There was realignment as well as remobilization.

The South survived the remobilization without much change. Turnouts remained at low levels. The exceptions were that what was left of the Republican party in the South diminished further, and Democratic politicians from the South could now exert much more influence than previously within the federal government. This latter change constrained the New Deal policies in many ways. The southern Democratic elite was much less interested in the redistributive and welfare aspects of the New Deal than were the northern Democrats. The tension between redistributive goals and accumulational ones has remained as a central divide within the Democratic "coalition" ever since.

Since 1940, and outside the special case of the South since 1948, decline in electoral participation has been across-the-board. Decline has been especially great since 1960, even given the fact that it has become much easier to vote in many parts of the country. The 46.9 percent *abstention* rate in the congressional election of 1980 was second only to the 47.3 percent of 1924. But it has been especially characteristic of the places and the people remobilized during the New Deal but "demobilized" since: the major industrial–urban centers, and the "lower half" of the social class structure. For example, in the 1980 congressional election, the two districts with the lowest participation rates were the 12th New York (Bedford–Stuyvesant) at 18.8 percent of estimated potential voters, and 22nd New York (Grand Concourse–Bronx) with a turnout of 21.8 percent. The two districts with the highest turnouts were wealthy suburban districts: the 10th Illinois (73.6 percent) in suburban Chicago, and the 3rd Minnesota (71.0 percent) in suburban Minneapolis. While the decline in electoral participation is broadly based, there are extreme cases. The greatest relative declines are concentrated in the central-city cores of the largest metropolitan areas, and in smaller but equally working-class settings elsewhere. Contrary to Verba and Nie's (1972:244) contention about low rates of participation in suburban areas, it is precisely in the wealthier suburbs that turnouts are now highest.

Burnham (1982) offers the plausible argument that rather than being "happy non-voters" with better things to do on election day than vote, the "party of the non-voters" are people who are attracted to parties that emphasize redistributive and welfare policies as the Democratic party did during its New Deal heyday. Today they are without such a vehicle. Since World War II the two major parties have both focused on accumulative strategies, with only periodic interest in redistributive policies. Neither can mobilize more than small fractions of the American potential electorate. The crisis of the Great Depression did do this by channelling nonvoters into a revitalized Democratic party. Since then, however, the needs of national economic management at home and the management of empire abroad have reduced the Democratic party's role and image as a surrog.te for a political movement relevant to the needs of nonvoters.

There is today, therefore, a crisis in electoral participation in the United States. By way of example, Ronald Reagan's popular mandate of 1980 rested upon only 28.0 percent of the potential electorate. This compares with the 28.3 percent Wendell Willkie obtained in 1940 while being trounced by Franklin D. Roosevelt, and the 40.2 percent that Giscard d'Estaing obtained of the French potential electorate in his *loss* to François Mitterand in 1981.

The American participation crisis reflects a deeper crisis of the state. Throughout the country, but especially in the older industrial–urban centers, a contradiction between the needs of capital accumulation and the elites that control it, on the one hand, and the needs of the state for coherence and legitimacy, on the other, is increasingly visible. Though the low new turnout among working-class people relieves the parties and the state of certain pressures, it also dangerously ignores the interests of the nonvoters. In many places in the contemporary United States, especially the industrial–urban centers, the legitimacy of the state is at issue as it has been perhaps at no time in the history of the United States. People are now *choosing* not to participate in national politics rather than being simply excluded from it. This is because accumulational strategies are necessarily confrontational. The financial and social burdens of reindustrializing and rearming America will fall disproportionately on those who "don't count" in the electoral calculus. Unless the Democratic party reverts to its New Deal form, those who presently don't count will continue to not vote. As they continue to be nonvoters the legitimacy of the state will continue to be called into question. Moreover, the explosive economic growth that worked in the 1890s to favor accumulational strategies at the expense of participation is a historical memory (Agnew 1987a). This is especially the case in the older industrial centers. As long as the economy "delivered," participation was a luxury; now that it does not, it is, from the viewpoint of legitimation, an absolute necessity. As Schattschneider (1960: 112) once put it

A greatly expanded popular base of political participation is the essential condition for public support of the government. This is the modern problem of democratic government. The price of support is participation. The choice is between participation and propaganda, between democratic and dictatorial ways of *changing consent into support, because consent is no longer enough* (emphasis in original).

Place and political parties

As electoral participation has declined, so has partisanship or party commitment. This is indicated above all by four trends: increased split-ticket voting; increased diversity of partisan swing across electoral districts; increased returns to incumbency irrespective of party; and, finally, the product of the other trends – partisan decay or disassociation between presidential and congressional politics. In each case the connection between place and the decline of partisanship is clear. Indeed, these trends all point to an increasing disintegration and denationalization of parties. National elections, especially congressional ones, are more and more *local* rather than national contests.

Burnham (1982) and others (Kleppner 1982, Wattenberg 1984) have argued that partisan decomposition has been the dominant feature of national politics in the United States since 1896. Beginning at that time, it increased through the 1920s, reversed temporarily during the New Deal era, and then returned to the 1920s level from the middle 1960s through until the present. It is a major concomitant *and* cause of the decline in electoral participation noted above. There are various ways of capturing the extent of "partisan decay." One is the extent of split-ticket voting, or voting for different offices for candidates who belong to different parties. Burnham (1985 : 235) has measured this using the results of elections in (a) all states outside the South with the congressional election concurrent with the presidential election, (b) the thirty-two states in which the senatorial elections were concurrent with the presidential election, and (c) the thirteen states in which gubernatorial elections coincided in 1984 with the presidential contest (Table 12.3). The stronger the partisanship the closer to 1.0 the r^2 will be between the voting for President and voting in the other contests. The weaker the partisanship, the closer to zero the r^2 will be.

The trend is unmistakable. From 1896–1900, when party was correlated highly across all contests, the r^2s move first downward (1900–28), then upward (1932–48), and then downward again (1952–84). But never have the correlations across all contests been as low as they are today. Even though 1984 showed an increase in the relationship between the presidential and congressional votes, it is still of a low order. Other relationships are so low that there is scarcely any positive relationship at all.

Table 12.3 The decline of partisanship: r^2 by state for president vs. US Senate, US House, governor, 1896–1984 (percent Democratic of the two-party vote).

Year	US Senate	US House	Governor
1896		0.941	0.963
1900		0.970	0.941
1904		0.798	0.726
1908		0.769	0.794
1912		0.610	0.891
1916	0.835	0.475	0.606
1920	0.798	0.695	0.546
1924	0.549	0.265	0.877
1928	0.295	0.239	0.678
1932	0.606	0.317	0.685
1936	0.700	0.559	0.598
1940	0.763	0.657	0.613
1944	0.879	0.680	0.758
1948	0.813	0.598	0.884
1952	0.620	0.750	0.649
1956	0.723	0.503	0.555
1960	0.425	0.546	0.358
1964	0.136	0.388	0.149
1968	0.229	0.257	0.096
1972	0.012	0.125	0.170
1976	0.070	0.187	0.439
1980	0.236	0.064	0.003
1984	0.143	0.258	0.012

Explanations and definitions
(1) Partisan percentages: Percentage Democratic of two-party vote, 1896–1908, 1916–20, 1928–44, 1952–84; Percentage Democratic of three-party vote, 1912; Percentage Democratic and Progressive of three-party vote, 1924; Percentage Democratic + Progressive + States' Rights Democratic of total vote, 1948.
(2) States with uncontested Senate or (for whole state) House elections omitted. For Senate, only elections for full terms are included. (Number of states ranges between 27 and 33.)
(3) The base of House elections are the contiguous 48 states less the 11 states of the ex-Confederacy; also less any other state where, statewide, there was no major-party opposition in any election. (Number ranges from 34 to 37.)
(4) The bases of gubernatorial elections are those 13 states in which, in 1984, presidential and gubernatorial elections occurred simultaneously. (N in 1896 = 12; thereafter 13.)
Source: Burnham (1985 : 235).

The rate at which the decoupling of party from national contests for different offices has been achieved has varied from place to place. In the period 1948–62 when the recent downward trend in partisanship took off nationally, there were places where it was less marked. In Michigan and Pennsylvania, for example, there was much less split-ticket voting than in Ohio and New York. This may well have been because the New Deal coalition and its policies remained both better organized and more relevant to the needs of its electorate in the former than in the latter. In certain other states, Oklahoma, for example, split-ticket voting has always been at relatively high levels. The recent trend there is an intensification of an old one, whereas elsewhere partisan decomposition is very much a product of either a trend that set in during the 1920s (Ohio) or since the 1960s (Pennsylvania) (Burnham 1982 : 31–44).

A second indicator of partisan decomposition is an increased diversity of swing across electoral districts. This phenomenon was referred to in Chapter 6 during the discussion of the nationalization thesis. Figure 6.7 shows, among other trends, the tendency for "swings" in votes between parties to be more heterogeneous recently (1972–4) than in the past (1862–4).

A congressional election is, at least in theory, a national event. This is the rationale behind the nationalization thesis of Stokes (1969). But the immense range of partisan swings suggests that a host of factors peculiar to each contest rather than national "party" swing are a major feature of recent congressional elections. Locality rather than party is the new byword for congressional candidates.

A third and associated measure of partisan decay is the increased role of incumbency in national elections. Miller and Stokes (1966) once pointed out that many voters know very little and few voters know very much about congressional candidates. This gives an edge to an incumbent because he has more visibility in his district than a challenger. Irrespective of partisanship, therefore, an incumbent has an advantage over a challenger (Fenno 1978). When added to by the favoritism extended to incumbents by the money raisers and campaign financiers looking out for their interests in Washington, this gives incumbents a major boost (Burnham 1985). Over the years there has been a general decline in electoral competitiveness as incumbency has become a major predictor of electoral outcomes. Since about 1956–60, incumbent members of Congress of both political parties have become increasingly invulnerable to defeat, regardless of the situation of their party's candidates for other offices in the same district (Tufte 1973). As the presidential elections of 1972 and 1984 demonstrated, congressional incumbents have become more or less insulated from the negative effects of landslides favorable to presidential candidates of the other party.

The higher visibility and resource-base of incumbents are undoubtedly of importance in explaining the power of incumbency. But they are by no means sufficient. Incumbents have always had such advantages. Why should incumbency have become so much more important after 1956–60? Certainly the changing "political technology" of campaigning has contributed heavily – congressional candidates selling themselves as district "troubleshooters," the growth of campaign advertising, and the increased focus on candidates as "personalities" rather than party representatives. Altogether these constitute what Blumenthal (1982) calls the "permanent campaign." With an election every other year, congressional candidates are

now full-time campaigners on their own behalf as much as party representation in Washington (Denzau *et al.* 1985).

However, the massive shift in favor of incumbency is also part of the
general decline of parties as "cue givers" for voting and agents of nationalization. Party organizations at the district level are now extremely weak, even
if more widespread because of the spread of two-party systems (Schlesinger
1985). They no longer provide a link between the voters and the politicians
(White 1983, Ware 1985).

The upshot of the decay of party in national elections is that though it
demonstrates the importance, indeed the increased importance in many
respects, of locality or place in national contests, it also cannot fail to have
dramatic consequences for the *center* of the political system as well. In
particular, the dissolution of the "partisan nexus" that once linked the
Presidency to Congress involves a major reinforcement of the separation of
powers. From 1964 to 1984, the Presidency and the Congress were held by
opposing parties over half the time; and from 1981 until 1984, Congress itself
was divided. This situation makes it difficult for government to follow
long-range objectives, and undermines public faith in government capabili-
ties. As one party wins the Presidency and the other the Congress, and the
Congress itself dissolves into representatives of places rather than parties, the
integration of the country is called into serious question.

Conclusion

Nation building and state integration depend critically upon popular mobilization for conflict management and legitimation. Political parties are important agents of these tasks. Clinton Rossiter (1960 : 24) has noted that:

> It has been [the parties] historic mission to hold the line against some of
> the most powerful centrifugal forces in American society ... If the
> parties continue to do their political, social and historical tasks with
> modest effectiveness, we need have few qualms about moving into the
> future with our pluralized form of government.

But such confidence has been seriously eroded by the decline in electoral
participation and increasing partisan decomposition. What is the likely
outcome of these trends? While the future cannot be predicted, a number of
possible outcomes can be explored.

One possibility is a permanent realignment. The prospects for this
scenario depend critically on the success of the present (1986) Republican
administration in realizing its objectives to the extent that the Republicans
become the equivalent of the Democrats at the time of the New Deal.
Though this is a possibility, it ignores the reality of a program that stands
only a slim chance of realization in present global economic conditions

(Agnew 1987a). It also ignores the shifts in capital investment and federal spending inside the United States that will be necessary for the realization of an American reindustrialization, and the consequent costs as well as benefits this will endow unequally in different parts of the country. The New Deal was at least an *integrating* coalition and program if nothing else. The present strategy is almost the opposite.

A second possibility is that if realignment fails, some kind of populist-fascism will enter the political stage to replace the present deadlock and drift. The current New Right, as Phillips (1982) points out, contains elements who are descendents of the old 19th century southern populist groups and who favor economic growth, economic nationalism, and the use of government to suppress forms of behavior they consider immoral at the expense of free-market ideology. However, as Phillips (1978) himself pointed out previously, the "balkanization" of America is now so developed that the possibility of a fascist or Nazi-type regime integrating these diverse interests is remote.

This leads to the third, and final, possibility. This is that low levels of participation and dealignment will continue. Beck (1984 : 265) argues that the dealignment that began in the 1960s may now be institutionalized and as such difficult to reverse. At least 30 or 40 years may be necessary for the "antipartisan" generation to disappear. Of course, they may leave a nonpartisan legacy behind with their offspring. To Burnham (1985 : 252-3), commenting on the 1984 election, the prospect is for a continued decay of party–voter links, and a continued role for "permanent campaigns" in increasingly disconnected *local* contests. The point of light in all this for the partisans of place is that it might lead us beyond the imperatives of liberal capitalism and the "imperial way of life" to a conception of politics in which the state is no longer the fundamental element of political life. The recent literature on American urban social movements (mainly neighborhood groups) points out that the growth in these movements is directly related to the decay of "conventional" politics (e.g., Katznelson 1981, Castells 1984). What it often misses is the empowerment of place through "partisan decay" that urban social movements now tap into. However, it does portray the potential for a revitalized political life from the bottom up, so to speak, that these movements represent. Or, as Williams (1981 : 225) expresses it,

> We all of us, here and everywhere, are in a transition period that offers us the opportunity to imagine and act upon a way to move on beyond global imperialism to regional communities. Away from the kind of interdependence programmed by the computers of the multinational corporations to the kind of dialogue that is the substance of a neighborhood.

The objective of Chapters 10 through 12 was to illustrate the efficacy of the place perspective by reference to the American case. As with Chapters 7 through 9 on Scottish politics, the purpose was not to provide an exhaustive survey of American political behavior. From three viewpoints – the United States as a whole, four places in particular, and the linkage provided by political parties between place and the state – an argument has been presented for examining American politics from the geographical perspective laid out in Chapter 3. Much of the information used in the "American" chapters is drawn from the work of others, particularly Bensel and Burnham. But, by and large, these authors have not made the intellectual "leap" to the position advocated here, even though their empirical work is often complementary to it. However, the significance of place has been difficult to acknowledge in an intellectual culture dominated by the nationalization thesis and the cult of the state. Only when the power of the state is viewed as contingent rather than complete can the power of place be admitted. Most students of American politics have not been prepared to go this far even when their "data" might suggest it. Evidently, facts are not enough.

13   Conclusion

This book has dual origins: as a response to the failure of political sociology to deal satisfactorily with contemporary events, such as the development of Scottish nationalism; and as a response to the problems posed to political sociology by the challenge of "contextual" social theory. The content and organization of the book represent these dual origins. The early chapters focus on some major conceptual questions posed by a reading of contemporary political sociology and the answers provided by an alternative perspective. The later chapters take up certain empirical themes in the context of first Scottish and then American politics.

The book appears at a time when there is a revival of interest in "contextual" social theory: achieving a merger of sociological and geographical "imaginations." Combining sociological and geographical imaginations has long been a key problem in social science, albeit neglected. The debate between the geographer Vidal de la Blache and the sociologist Durkheim in turn-of-the-century France was partly a clash over the *priority* to be given one or the other. The geographical imagination is a concrete and descriptive one, concerned with determining the nature of and classifying places. But the sociological imagination aspires to explanation of people and places in terms of social process abstractly, and often nationally, construed. Thus, Vidal emphasized the gathering of "facts" about places, whereas Durkheim stressed the morphology of society rather than landscape. "Social facts" rather than "place facts" defined the territory of the sociologist. Thereafter, integrating the two imaginations became not only difficult, but because of institutionalized methodological and conceptual differences, well-nigh impossible.

The past few years have seen a resurgence of interest in the possibility and importance of bringing together geographical and sociological imaginations. Some geographers have been particularly active in attempting to achieve a "marriage" of the two minds. The work of Harvey (1973), Gregory (1978, 1982) and Pred (1983, 1984) is exemplary. Many geographers, however, remain attached to a "spatial separatist" image of their discipline. This is so despite the problems that arise when "space" or distance is accorded a singular role as a "cause" of human activity (Sack 1973, 1974; N. Smith 1981). Specifically, the use of spatial concepts carries the inherent danger of hypostatizing space – making incorrect inference from geographical pattern to sociological process, and reading permanence into historically constituted and dynamic geographical relationships (Agnew 1982). Other social scientists seem to be intent on making the same mistakes.

This is particularly true of proponents of "world-system" and underdevelopment theories, at least in some of their writings (e.g., Chase-Dunn & Rubinson 1977, Wallerstein 1979, Bergesen 1980). But recent work in social theory and philosophy, expressed for example in the work of Foucault (1980) and Giddens (1979, 1981), represents a movement towards a meeting, if not a marriage, between the two imaginations. These writers emphasize the embeddedness of social relations in space–time contexts rather than their mutual exclusivity or singular determinism.

As yet, however, developments in social theory have not filtered down into the practice of social science. This is clearly the case in political sociology, where sociological and, now more frequently, psychological explanations are predominant (Ch. 2; Rosenberg 1985). But the conventional modes of theorizing are approaching a crisis point. Wedded as they are to a "nationalization" conception of political change and behavior, they cannot adequately cope with evidence that contradicts it. That evidence is now legion.

Chapters 6 through 12 of this book offer considerable evidence both *against* the nationalization thesis in a variety of contexts, and *for* the place perspective outlined in Chapter 3, especially in terms of American and Scottish politics. Alternatives such as the neighborhood effect, uneven development, and dealignment were considered but rejected as insufficient. Rather, an explanation couched in terms of place-specific economic growth and social change and associated political identities is preferred.

A major purpose of this book, therefore, is to provide a geographical–sociological perspective from which to make sense of the failure of the nationalization thesis and respond to the intellectual crisis facing political sociology. But the book also begins from the premise that political behavior can best be explained contextually, in terms of the social contexts designated by the word "place." Given that the prime objective of political sociology is causal explanation of political activities, an emphasis on place underlines a priority to causal validity rather than generalization. It is *through* places, so to speak, that social causes "produce" behavior. But it is also *in* places that human agency produces and reproduces social causes. Or, to use the vocabulary of power, it is in places that the power to act as human agents is both channeled and realized within limits laid down by locally dominant practices within which are incorporated extralocal influences or powers.

What has been needed is a social theory that provides a linkage between, on the one hand, locally structured microsociological arrangements within which agency is realized (work, home, school, church, etc.), and, on the other hand, the "structurally determined" limits set by the macro-order through restricting, directing, and obscuring agency. Such a theory has been provided here in terms of the three "faces" or dimensions of place: locale, location, and sense of place. Locale and sense of place describe, respectively, the objective and subjective dimensions of local social arrangements.

Location refers to the impact of the macro-order, to the fact that a single place is one among many and subject to influence from these others, and that the social life of a place is also part of the life of a state and the world-economy. Taken together they constitute the defining elements of a place as a historically constituted social context for political and other forms of social behavior.

The major thrust of the empirical chapters was towards illustrating the efficacy of this theory. In the Scottish chapters, the major conclusion is that support for the various political parties, both British and Scottish, cannot be explained satisfactorily without recourse to the conception of place outlined in Chapter 3. In the American chapters a rather more broad-based empirical analysis gives strong support to the place perspective in the homeland of the nationalization thesis. Indeed, some of the material used in the American chapters suggests that place is now perhaps more *clearly* important for the structuration of political behavior than it was in the past.

As I have worked on this book, colleagues have offered a number of criticisms. Some of these are dealt with in the body of the text. But the conclusion offers an opportunity to admit, describe, and confront criticisms that may have persisted in spite of efforts to answer them or that have not yet been answered head-on.

One criticism of the position taken in this book is that it is a massive instance of the so-called "ecological fallacy." This was originally a statistical argument, if one with ontological overtones. It states that an ecological correlation is not equal to a corresponding "individual" correlation, and that in explaining individual behavior individual correlations are preferable (Robinson 1950). But it is clear that ecological correlations *are* relevant to ecological interpretations of individual behavior and not merely substitutes for individual ones (Menzel 1950). This criticism, therefore, is only effective when ecological data are used solely to make inferences about individual-level or *psychological* causes of individual behavior (Allardt 1969). If one's ontology is not individualistic but social, the criticism loses much of its power (Wald 1983). However, given the reference to "agency," I would be remiss in not pointing out that a major deficiency in the chapters on specific places (Chs. 8 & 11) is a lack of *direct* information from individuals about their conceptions of politics and place. A major priority for future research should be the incorporation of individual "reason-giving" accounts into place vignettes.

Another criticism is of the nonindividualist ontology itself. According to the "modernization" theory of social change that has dominated Western social science in recent times, place is a locus for significant social relations in traditional but not in modern societies. "Becoming modern" involves casting off ties to place (in work, recreation, and sense of identity) and adopting an "achievement-oriented" or "class-conscious" self that is place-less. This is a *natural* process, independent of human will and beyond human control. Consequently, to focus on places in modern societies is fetishistic, since place has no modern meaning except as the incidental location of society-wide activities or the mere intersection of map coordinates.

To the extent that place has its defenders, they are often political conservatives. They see a golden age of place in the past, but regret its demise and decry the "culture of modernism" that has replaced it. In particular, they yearn for the hierarchical social relations and secondary social institutions (family, church, etc.) that they associate with communal–place-based social life and which have been "eclipsed" by the advent of modernism (Nisbet 1953, Bell 1977). Modernization theory, then, leads towards an emphasis on individuals and their troubles, whereas the critique of modernism stresses the *virtue* of social hierarchy and place-based communalism. However, both agree that the "era of place" has passed. The only suitable ontology for today, they argue, is individualist.

In Chapter 5 it was argued that this viewpoint is deficient. At the crux of the "devaluation of place" is a confusion between place and community. Community has acquired two distinct meanings: a morally valued way of life, and the constituting of social relations in a discrete geographical setting. But rather than distinguishing these two connotations, modern social science has conflated them. In particular, a specific set of social relations – those of a morally valued way of life – have transcended the generic sense of community-as-place. This emphasis has discounted the possibility of *new* forms of community, and more critically in the present connection, prevented the possibility of seeing society-in-place.

One important correlate and cause of the devaluation of place is the definition of society as coterminous with the boundaries of existing states. The tendency to accept the modern nation-state as "natural," and indeed progressive, has led to a "state-centered" social science and political sociology. It is significant that the modern social science disciplines and their central concepts developed at a time and in those states caught up in the process of "nation building." Alternatives were ignored or crushed.

In the field of geography, for example, the statist political geography of Mackinder rather than the antistatist political geography of Reclus and Kropotkin was institutionalized. The latter's political geography, focusing on local socioterritorial contexts and the impacts of the new states and industrial capitalism on them, brought with it *political* commitments and an antistatist ideology that were dangerous to the emerging orthodoxy (MacLaughlin 1987). To this day, the "alternative" political geography of Reclus and Kropotkin has been ignored in conventional histories of geographic thought (e.g., Freeman 1961, James & Martin 1981). Yet it rested on a "classic" geographical premise: assume away all geographical barriers presented by our present mode of living such as scarcity and alienation and it still will be the case that we inhabit space, and space inherently creates social relations of presence (or proximity) and social relations of absence (or distance). This points, as Connolly (1977) reminds those blinded by the fantasy of a "spaceless society," to the need for *political* activities and institutions even in a "socialist order" in which, at least on non-Stalinist assumptions, there will be competing bases for identification, and hence a need for political resolution. Indeed, if one sees an identity between socialism and "strong democracy" or greater popular participation, the continuing significance of place is a plus rather than a minus. For it is parochial participation and local activity that can transform passive consumers of elite policies into active citizens, as witnessed to by the growth of "community politics" in American cities in the late 1960s.

A third and final criticism is that "place" may well *help* explain political life but cannot be the whole story. In one sense this is on the mark. As Chapters 3, 7, and 10 make clear, places cannot be understood in isolation from one another, or from structures that they produce and reproduce but which are "larger" than any one of them. The "national" political institutions and global economic processes to which local political life is often oriented and which influence local practices form a particularly important backdrop to place-specific activities. In another sense, however, this criticism is misleading. It rests on the assumption that place is important *solely* as a separable local component to political behavior in general. Place is yet another dummy variable for a multiple regression equation (e.g., McAllister 1987). It is one *scale* among several. Add them all up and you have a complete explanation (e.g., Urwin 1980). This is much the same reasoning as lies behind neighborhood effect explanations of uniform national swing in British elections (e.g., Butler & Stokes 1969).

But the basic challenge of microsociology upon which the place perspective is based is that political order is produced and reproduced through microsociological routines (locale and sense of place). Whatever the specific nature of power relationships, they cannot be separated from the realm of action and everyday practices. The macro-order (location) is represented in the routines and practices of people in places. It is itself the product of the consequences, intended and unintended, of previous and present-day micro-situations and power relationships, especially those of firms and governments with a wide and intensive "reach" into other places. Though often transparent to the actor, the effects of the macro-order are real enough (Layder 1985). So much so that one should be careful not to allow as much scope for "free" local action as some advocates of contextual social theory may be inclined (e.g., Clark 1985). If, as Bottomore (1984 : 75) contends,

> the real problem [in social science] is to formulate a conception of social structure which does justice to the elements of regularity and order in social life, while not neglecting the flow of historical action by individuals and social groups which sustains, recreates, revises, or disrupts this order

then the concept of place as developed and applied here may just provide one appropriate response.

References

Agnew, J. A. 1978. Neighbourhood schools and social mix: is comprehensive education an incentive for residential segregation? *Area* **10**, 318–20.

Agnew, J. A. 1981a. Political regionalism and Scottish nationalism in Gaelic Scotland. *Can. Rev. Studs. Nat.* **8**, 115–29.

Agnew, J. A. 1981b. Structural and dialectical theories of political regionalism. In *Political studies from spatial perspectives: Anglo-American essays in political geography*, A. Burnett & P. J. Taylor (eds.). Chichester: Wiley.

Agnew, J. A. 1981c. Home ownership and identity in capitalist societies. In *Housing and identity: cross-cultural perspectives*, J. S. Duncan (ed.). London: Croom Helm.

Agnew, J. A. 1982. Sociologizing the geographical imagination: spatial concepts in the world-system perspective. *Pol. Geog. Q.* **1**, 159–66.

Agnew, J. A. 1983. An excess of 'national exceptionalism': towards a new political geography of American foreign policy. *Pol. Geog. Q.* **2**, 151–66.

Agnew, J. A. 1984a. Place and political behaviour: The geography of Scottish nationalism. *Pol. Geog. Q.* **3**, 191–206.

Agnew, J. A. 1984b. Devaluing place: 'people prosperity' versus 'place prosperity' and regional planning. *Soc. and Space* **2**, 35–45.

Agnew, J. A. 1987a. *The United States in the world-economy: a regional geography.* Cambridge: Cambridge University Press.

Agnew, J. A. 1987b. Place anyone? comments on the papers of McAllister and Johnston. *Pol. Geog. Q.* **6**, 39–40.

Agnew, J. A. & J. S. Duncan 1981. The transfer of ideas into Anglo-American human geography. *Prog. Hum. Geog.* **5**, 42–57.

Agulhon, M. 1950. L'Opinion politique dans une commune de banlieue sous la troisième république: Bobigny de 1850 à 1914. In *Études sur la banlieue de Paris*. Paris: Armand Colin, Cah. Fond. Nat. Sci. Pol. 12.

Agulhon, M. 1983. *The Republic in the village: the people of the Var from the French Revolution to the Second Republic.* Cambridge: Cambridge University Press.

Aiken, M. & G. Martinotti 1982. Sistema urbano, governo della città e giunte di sinistra nei grandi comuni italiani. *Quad. Sociol.* **2–4**, 177–248.

Alford, R. R. 1963. *Party and society: the Anglo-American democracies.* Chicago: Rand McNally.

Allardt, E. 1969. Aggregate analysis: the problem of its informative value. In *Social ecology*, M. Dogan & S. Rokkan (eds.). Cambridge, Mass.: MIT Press.

Allum, P. 1974. The Neapolitan politicians: a collective portrait. In *The politics and society reader*, I. Katznelson (ed.). New York: David McKay.

Almond, G. & J. S. Coleman (eds.) 1960. *The politics of the developing areas.* Princeton N.J.: Princeton University Press.

Almond, G. & S. Verba 1963. *The civic culture.* Boston: Little, Brown.

Althusser, L. 1970. *For Marx*, New York: Vintage.

Andersen, K. 1979. *The creation of a Democratic majority, 1928–1936.* Chicago: University of Chicago Press.

Andrews, W. G. 1974. Social change and electoral politics in Britain: a case study of Basingstoke, 1964 and 1974. *Pol. Studs* **22**, 324–36.

Appleby, J. O. 1978. *Economic thought and ideology in seventeenth-century England.* Princeton, N.J.: Princeton University Press.

Archer, C. & P. J. Taylor 1981. *Section and party: a political geography of American presidential elections, from Andrew Jackson to Ronald Reagan*. Chichester: Wiley.

Archer, J. C., G. T. Murauskas, F. M. Shelley, E. R. White & P. J. Taylor 1985. Counties, states, sections and parties in the 1984 presidential election. *Prof. Geog.* **37**, 279–87.

Arlacchi, P. 1983. *Mafia, peasants and great estates: society in traditional Calabria*. Cambridge: Cambridge University Press.

Aron, R. 1955. Réflexions sur la politique et la science politique française. *Rev. française sci. pol.* **5**, 5–20.

Asher, H. 1980. *Presidential elections and American politics*, 2nd edn. Homewood, Ill.: Dorsey.

Ashford, D. 1975. Resources, spending and party politics in British local government. *Admin. and Soc.* **7**, 74–94.

Aymard, M. 1972. The *Annales* and French historiography (1929–1972). *J. Eur. Econ. Hist.* **1**, 491–511.

Baker, A. R. H. 1984. Reflections on the relations of historical geography and the *Annales* school of history. In *Explorations in Historical Geography*, A. R. H. Baker & D. Gregory (eds.). Cambridge: Cambridge University Press.

Baker, K. L., R. J. Dalton & K. Hildebrandt 1981. *Germany transformed*. Cambridge, Mass.: Harvard University Press.

Balsom, D. 1979. *The nature and distribution of support for Plaid Cymru*. Glasgow: University of Strathclyde, Stud. Pub. Pol. 36.

Baltzell, D. 1979. *Puritan Boston, Quaker Philadelphia*. Philadelphia: University of Pennsylvania Press.

Banfield, E. C. 1965. *Big city politics*. New York: Random House.

Barnard, F. 1969. *Herder on social and political culture*. Cambridge: Cambridge University Press.

Barth, F. (ed.) 1963. *The role of the entrepreneur in social change in northern Norway*. Oslo: Universitetsforlaget.

Barth, F. (ed.) 1969. *Ethnic groups and boundaries*. Boston: Little, Brown.

Bastide, R. 1966. Y a-t-il une crise de la psychologie des peuples? *Rev. psych. peuples* **21**, 8–20.

Bealey, F. & J. Sewel 1981. *The politics of independence*. Aberdeen: Aberdeen University Press.

Beck, P. A. 1977. Partisan dealignment in the postwar South. *Amer. Pol. Sci. Rev.* **71**, 477–96.

Beck, P. A. 1984. The dealignment era in America. In *Electoral change in advanced industrial democracies: realignment or dealignment?* R. J. Dalton, S. C. Flanagan & P. A. Beck (eds.). Princeton, N.J.: Princeton University Press.

Bell, C. & H. Newby 1971. *Community studies: an introduction to the sociology of the local community*. London: Allen & Unwin.

Bell, D. 1976. *The cultural contradictions of capitalism*. New York: Basic Books.

Bell, D. H. 1984. Working-class culture and Fascism in an Italian industrial town, 1918–22. *Soc. Hist.* **9**, 1–24.

Bender, T. 1978. *Community and social change in America*. New Brunswick, N.J.: Rutgers University Press.

Bennett, S. & C. Earle 1983. Socialism in America: a geographical interpretation of its failure. *Pol. Geog. Q.* **2**, 31–55.

Bensel, R. F. 1984. *Sectionalism and American political development, 1880–1980*. Madison, Wis.: University of Wisconsin Press.

Benson, J. 1980. *British coalminers in the nineteenth century*. New York: Holmes & Meier.

Berelson, B., P. Lazarsfeld & W. H. McPhee 1954. *Voting*. Chicago: University of Chicago Press.

Berger, P. & T. Luckmann 1966. *The social construction of reality*. New York: Doubleday.

Berger, S. 1972. *Peasants against politics: rural organization in Brittany, 1911–1967*. Cambridge, Mass.: Harvard University Press.

Bergesen, A. 1980. From utilitarianism to globology: the shift from the individual to the world as a whole as the primordial unit of analysis. In *Studies in the modern world-system*, A. Bergesen (ed.). New York: Academic Press.

Bernard, J. 1973. *The sociology of community*. Glenview, Ill.: Scott, Foresman.

Berrington, H. 1984. Change in British politics: an introduction. In *Change in British politics*, H. Berrington (ed.). London: Frank Cass.

Berry, B. J. L. *et al.* 1976. Attitudes towards integration: the role of status in community response to racial change. In *The changing face of the suburbs*, B. Schwartz (ed.). Chicago: University of Chicago Press.

Bhaskar, R. 1979. *The possibility of naturalism*. Hassocks, England: Harvester Press.

Bhaskar, R. 1983. Beef, structure and place: notes from a critical naturalist perspective. *J. Theory Soc. Behav.* **13**, 81–95.

Birnbaum, P. 1970. *Sociologie de Tocqueville*. Paris: Presses Universitaires de la France.

Blackbourn, D. 1980. *Class, religion, and local politics in Wilhelmine Germany: the Centre party in Württemburg before 1914*. New Haven, Conn.: Yale University Press.

Blewett, N. 1965. The franchise in the United Kingdom, 1885–1918. *Past and Present* **32**, 27–56.

Blondel, J. 1963. *Voters, parties, and leaders*. London: Penguin.

Bluestone, B. & B. Harrison 1982. *The deindustrialization of America*. New York: Basic Books.

Blumenthal, S. 1982. *The permanent campaign*. New York: Simon & Schuster.

Blumler, J. G. & D. McQuail 1969. *Television in politics: its uses and influence*. Chicago: University of Chicago Press.

Bochel, J. M. & D. T. Denver 1975. *The Scottish local government elections 1974*. Edinburgh: Scottish Academic Press.

Bochel, J. M. & D. T. Denver 1978. *The Scottish regional elections 1978*. Dundee: University of Dundee, Department of Political Science.

Bochel, J. M. & D. T. Denver 1980. *The Scottish district elections 1980*. Dundee: University of Dundee, Department of Political Science.

Bochel, J. M. & D. T. Denver 1982. *Scottish regional elections 1982*. Dundee: University of Dundee, Department of Political Science.

Bochel, J. M. & D. T. Denver 1983. *The new Scottish constituencies: a guide and an analysis*. Dundee: University of Dundee, Department of Political Science.

Bochel, J. M., D. T. Denver & A. J. Macartney (eds.) 1981. *The referendum experience: Scotland 1979*. Aberdeen: Aberdeen University Press.

Bodman, A. R. 1982. Measuring political change. *Env. Plann.* **A 14**, 33–48.

Bodman, A. R. 1983. The neighbourhood effect: A test of the Butler–Stokes model. *Br. J. Pol. Sci.* **12**, 124–31.

Bodman, A. R. 1985. Regional trends in electoral support in Britain, 1950–1983. *Prof. Geog.* **37**, 288–95.

Bogdanor, V. 1983. *Multi-party politics and the constitution*. Cambridge: Cambridge University Press.

Bottomore, T. 1979. *Political sociology*. New York: Harper & Row.

Bottomore, T. 1984. *Sociology and socialism*. Brighton: Wheatsheaf.

Bourdieu, P. 1977. *Outline of a theory of practice*. Cambridge: Cambridge University Press.

Bowler, I. R. 1975. Regional variations in Scottish agricultural trends. *Scot. Geog. Mag.* **91**, 114–22.

Brittan, A. 1973. *Meanings and situations*. London: Routledge & Kegan Paul.

Brand, J. 1978. *The national movement in Scotland*. London: Routledge & Kegan Paul.

Brand, J. & D. McCrone 1975. The SNP: from protest to nationalism. *New Society* November 20, 16–18.

Brand, J., D. McLean & W. Miller 1983. The birth and death of a three-party system: Scotland in the seventies. *Br. J. Pol. Sci.* **13**, 463–88.

Brenner, R. 1977. The origins of capitalist development: a critique of neo-Smithian marxism. *New Left Rev.* **104**, 25–92.

Brookfield, H. 1978. Third world development. *Progr. Hum. Geog.* **2**, 121–32.

Browne, E. C. & L. L. Vertz 1983. An old people in a new state: the problem of national integration in West Germany. *Comp. Pol.* **16**, 85–95.

Browne, L. 1983. Can high-tech save the Great Lake states? *New Engl. Econ. Rev.* November–December, 19–33.

Browne, L. 1984. How different are regional wages? a second look, *New Engl. Econ. Rev.* March–April, 40–7.

Browne, L. & J. S. Hekman 1981. New England's economy in the 1980s. *New Engl. Econ. Rev.* January–February, 5–16.

Brusa, C. 1983. *Geografia elettorale nell 'italia del dopoguerra*. Milan: Unicopli.

Brusa, C. 1984a. *Geografia del potere politico in Italia*. Milan: Unicopli.

Brusa, C. 1984b. *Geografia elettorale nell 'italia del dopoguerra: edizione aggiornata ai risultati delle elezioni politiche 1983*. Milan: Unicopli.

Brustein, W. 1981. A regional mode-of-production analysis of political behavior: the cases of Mediterranean and western France. *Pol. and Soc.* **10** (4), 355–98.

Budge, I. & D. Urwin 1966. *Scottish political behaviour: a case study in British homogeneity*. London: Longman.

Budge, I., I. Crewe & D. Farlie (eds.) 1976. *Party identification and beyond*. Chichester: Wiley.

Bullock, A. 1964. *Hitler: a study in tyranny*, 2nd edn. New York: Harper & Row.

Bunge, W. & R. Bordessa 1975. *The Canadian alternative: survival, expeditions and urban change*. Toronto: York University Department of Geography.

Burman, P. 1979. Variations on a dialectical theme. *Phil. Soc. Sci.* **9**, 357–75.

Burnham, W. D. 1965. The changing shape of the American political universe. *Am. Pol. Sci. Rev.* **59**, 7–28.

Burnham, W. D. 1981a. Toward confrontation? In *Party coalitions in the 1980s*, S. M. Lipset (ed.). San Francisco: Institute for Contemporary Studies.

Burnham, W. D. 1981b. The 1980 earthquake: realignment, reaction, or what? In *The hidden election: politics and economics in the 1980 presidential campaign*, T. Ferguson & J. Rogers (eds.). New York: Pantheon.

Burnham, W. D. 1982. *The current crisis in American politics*. New York: Oxford University Press.

Burnham, W. D. 1984. Parties and political modernization. In *Political parties and the modern state*, R. L. McCormick (ed.). New Brunswick, N.J.: Rutgers University Press.

Burnham, W. D. 1985. The 1984 election and the future of American politics. In *Election '84: landslide without a mandate?* E. Sandoz & C. V. Crabb Jr. (eds.). New York: Mentor.

Business Week 1984. Latin America's woes are casting a pall over Miami. April 30, 134–43.

Butler, D. & D. Stokes 1969. *Political change in Britain: forces shaping electoral choice*. London: Macmillan.

Butler, D. & D. Kavanagh 1975. *The British general election of October 1974*. London: Macmillan.

Cahnman, W. J. 1977. Töennies in America. *Hist. and Theory* **16**, 147–67.

Calhoun, C. J. 1978. History, anthropology and the study of communities: some problems in Macfarlane's proposal. *Soc. Hist.* **3**, 363–73.

Calhoun, C. J. 1980. Community: toward a variable conceptualization for comparative research. *Soc. Hist.* **5**, 105–29.

Campbell, A., P. E. Converse, W. E. Miller & D. Stokes 1960. *The American voter*. New York: Wiley.

Campbell, A., P. E. Converse, W. E. Miller & D. Stokes 1966. *Elections and the political order*. New York: Wiley.

Campbell, R. H. 1980. *The rise and fall of Scottish industry, 1707–1939*. Edinburgh: Donald.

Capecchi, V. & Galli, G. 1969. Determinants of voting behavior in Italy: a linear causal model of analysis. In *Social ecology*, M. Dogan & S. Rokkan (eds.). Cambridge, Mass.: MIT Press.

Cardoza, A. L. 1982. *Agrarian elites and Italian fascism: the province of Bologna, 1901–1926*. Princeton N.J.: Princeton University Press.

Carr, E. H. 1961. *What is history?* New York: Knopf.

Carter, C. J. 1974. Some post-war changes in the industrial geography of the Clydeside conurbation. *Scott. Geog. Mag.* **90**, 14–27.

Carter, I. 1974. The Highlands of Scotland as an underdeveloped region. In *Sociology and Development*, E. deKadt & G. Williams (eds.). London: Tavistock.

Carter, I. 1976. Class and culture among farm servants in the northeast of Scotland, 1840–1914. In *Social class in Scotland: past and present*, A. A. MacLaren (ed.). Edinburgh: Donald.

Carty, R. K. 1981. *Party and parish pump: electoral politics in Ireland*. Waterloo, Canada: Wilfrid Laurier University Press.

Castells, M. 1984. *The city and the grassroots*. Berkeley, Calif.: University of California Press.

Caudill, H. M. 1983. *Theirs be the power: the moguls of eastern Kentucky*. Urbana, Ill.: University of Illinois Press.

Chapman, K. 1982. Energy production and use. In *The changing geography of the United Kingdom*, R. J. Johnston & J. C. Doornkamp (eds.). London: Methuen.

Chase-Dunn, C. & R. Rubinson 1977. Toward a structural perspective on the world-system. *Pol. and Soc.* **7**, 453–76.

Checkland, S. G. 1976. *The Upas tree: Glasgow 1875–1975, a study in growth and contraction*. Glasgow: University of Glasgow Press.

Cicourel, A. 1968. *The social organization of juvenile justice*. New York: Wiley.

Claggett, W., W. Flanigan & N. Zingale 1984. Nationalization of the American electorate. *Am. Pol. Sci. Rev.* **78**, 77–91.

Clark, G. L. 1984. A theory of local autonomy. *Ann. Assoc. Am. Geogs.* **74**, 195–208.

Clark, G. L. 1985. *Judges and the cities: interpreting local autonomy*. Chicago: University of Chicago Press.

Clark, T. N. 1973. *Community power and policy outputs*. Beverly Hills, Calif.: Sage.

Clark, T. N. 1975. The Irish ethic and the spirit of patronage. *Ethnicity* **2**, 15–33.

Clarke, M. G. & H. M. Drucker 1975. Our changing Scotland. In *Our changing Scotland*, M. G. Clarke & H. M. Drucker (eds.). Edinburgh: Edinburgh University Student Publications Board.

Claval, P. 1984. *Models of man in geography*. Syracuse, N.Y.: Syracuse University, Department of Geography Discussion Paper Series 79.

Cloos, G. & P. Cummins 1984. Economic upheaval in the Midwest. *Econ. Perspect.* January–February, 3–14.

Cohen, P. C. 1982. *A calculating people: the spread of numeracy in early America.* Chicago: University of Chicago Press.

Collingwood, R. G. 1948. *The idea of history.* London: Oxford University Press.

Collins, R. 1981. On the micro-foundations of macro-sociology. *Am. J. Soc.* **86**, 984–1014.

Connolly, W. 1977. A note on freedom under socialism. *Pol. Theory* **5**, 461–72.

Conradt, D. P. 1973. *The West German party system: an ecological analysis of social structure and voting behavior, 1961–1969.* Beverly Hills, Calif.: Sage.

Converse, P. 1972. Change in the American electorate. In *The human meaning of social change*, A. Campbell & P. Converse (eds.). New York: Russell Sage.

Corner, P. 1975. *Fascism in Ferrara, 1915–1922.* Oxford: Clarendon Press.

Coulin, C. 1978. French political science and regional diversity: a strategy of silence. *Ethn. Rac. Stud.* **1**, 80–99.

Cox, K. R. 1968. Suburbia and voting behavior in the London metropolitan area. *Ann. Assoc. Am. Geogs.* **58**, 111–27.

Cox, K. R. 1969a. The voting decision in a spatial context. *Prog. Geogr.* **1**, 81–117.

Cox, K. R. 1969b. On the utility and definition of regions in comparative political sociology. *Comp. Pol. Stud.* **2**, 68–98.

Cox, K. R. 1972. The neighborhood effect in urban voting response surfaces. In *Models of urban structure*, D. Sweet (ed.). Lexington, Mass.: Lexington.

Cox, K. R. 1978. Local interests and urban political processes in market societies. In *Urbanization and conflict in market societies*, K. R. Cox (ed.). Chicago: Maaroufa.

Cox, K. R. 1985. *The urban development process and the rise of the new spatial politics.* Unpubl. paper, Dept. of Geog., Ohio St. Univ., Columbus, Ohio.

Crampton, P. 1984. Spatial polarization of political representation in Great Britain: 1945–1979. *Geog.* **69**, 28–37.

Craig, F. W. S. 1969. *British parliamentary election results 1918–1949.* Glasgow: Political Reference Publications.

Craig, F. W. S. 1971. *British parliamentary election statistics, 1918–1970.* Chichester: Political Reference Publications.

Craig, F. W. S. 1975. *Minor parties at British parliamentary elections, 1885–1974.* London: Macmillan.

Crewe, I. & C. Payne 1971. Analyzing the census data. In *The British General Election of 1970*, D. Butler & M. Pinto-Duschinsky (eds.). London: Macmillan.

Crewe, I. & C. Payne 1976. Another game with nature: an ecological regression model of the two-party vote ratio in 1970. *Br. J. Pol. Sci.* **6**, 43–81.

Crewe, I., B. Särlvik & J. Alt 1977. Partisan dealignment in Britain. *Br. J. Pol. Sci.* **7**, 129–90.

Crocker, L. G. 1959. *An age of crisis: man and world in eighteenth-century French thought.* Baltimore: Johns Hopkins University Press.

Cunnison, J. & J. Gilfillan (eds.) 1958. *The third statistical account of Scotland: Glasgow.* Glasgow: Collins.

Dahmann, D. C. 1982. *Locals and cosmopolitans: patterns of spatial mobility during the transition from youth to early adulthood.* Chicago: University of Chicago, Department of Geography Research Paper No. 204.

Dalton, R. J., S. Flanagan & P. A. Beck (eds.) 1984. *Electoral change in advanced industrial democracies.* Princeton, N.J.: Princeton University Press.

Danielson, M. N. 1976. *The politics of exclusion.* New York: Columbia University Press.

Darragh, J. 1978. The Catholic population of Scotland 1878–1977. *Innes Rev.* **29**, 211–47.

Davis, M. 1981. The New Right's road to power. *New Left Rev.* **128**, 28–49.

Deans, P. 1978. *The evolution of economic ideas.* Cambridge: Cambridge University Press.

Denzau, A., W. Riker & K. Shepsle 1985. Farquharson and Fenno: sophisticated voting and home style. *Am. Pol. Sci. Rev.* **70**, 1117–34.

Derivry, D. & M. Dogan 1986. Religion, classe et politique en France: six types de relations causales. *Rev. française sci. pol.* **36**, 157–81.

Desai, M. 1986. Men and things. *Economica* **53**, 1–10.

Deutsch, K. 1953. *Nationalism and social communication.* Cambridge, Mass.: MIT Press.

Deutsch, K. 1966. *The nerves of government.* New York: Free Press.

Devine, D. J. 1972. *The political culture of the United States: the influence of member values on regime maintenance.* Boston: Little, Brown.

De Vito, A. P. 1979. Urban revitalization: the case of Detroit. *Urb. Concerns* May–June, 3–13.

Dicken, P. 1982. The industrial structure and the geography of manufacturing. In *The changing geography of the United Kingdom*, R. J. Johnston & J. C. Doornkamp (eds.). London: Methuen.

Dogan, M. & S. Rokkan (eds.) 1969. *Social ecology.* Cambridge, Mass.: MIT Press.

Domhoff, W. 1967. *Who rules America?* Englewood Cliffs, N.J.: Prentice-Hall.

Downs, A. 1957. *An economic theory of democracy.* New York: Harper & Row.

Drucker, H. M. 1978. *Breakaway: the Scottish Labour party.* Edinburgh: Edinburgh University Student Publications Board.

Drucker, H. M. & G. Brown 1980. *The politics of nationalism and devolution.* London: Longman.

Dulong, R. 1978. *Les régions, l'état et la societé locale.* Paris: Presses Universitaires de la France.

Dunbabin, J. P. D. 1974. *Rural discontent in nineteenth-century Britain.* New York: Holmes & Meier.

Dunbabin, J. P. D. 1980. British elections in the nineteenth and twentieth centuries, a regional approach. *Eng. Hist. Rev.* **375**, 241–67.

Dunleavy, P. 1977. Protest and quiescence in urban politics: a critique of some pluralist and structuralist myths. *Int. J. Urb. Reg. Res.* **1**, 193–218.

Dunleavy, P. 1979. The urban basis of political alignment: social class, domestic property ownership, and state intervention in consumption processes. *Br. J. Pol. Sci.* **9**, 409–43.

Dunleavy, P. 1980. The political implications of sectoral cleavages and the growth of state employment: Part 1, The analysis of production cleavages. Part 2, Cleavage structures and political alignment. *Pol. Studs.* **28** (3):364–83, (4):527–49.

Durkheim, E. 1904. Review of A. Allin *The basis of sociality. L'Année Sociologique* **7**, 185.

Durkheim, E. 1933. *De la division du travail social*, 2nd edn. Paris: Alcan.

Duverger, M. 1959. *Political parties: their organization and activity in the modern state.* London: Methuen.

Duverger, M. 1973. *La sociologie de la politique, eléments de science politique.* Paris: Presses Universitaires de France.

Dyer, M. 1981. Aberdeen and the Grampian region. In *The referendum experience – Scotland 1979*, J. M. Bochel, D. T. Denver & A. J. Macartney (eds.). Aberdeen: Aberdeen University Press.

Easton, D. 1953. *The political system.* New York: Knopf.

Effrat, A. (ed.) 1972. *Perspectives in political sociology*. Indianapolis: Bobbs-Merrill.

Elazar, D. J. 1972. *American federalism: a view from the states*. New York: Crowell.

Eller, R. D. 1982. *Miners, millhands, and mountaineers: industrialization of the Appalachian south, 1880–1930*. Knoxville, Tenn.: University of Tennessee Press.

Ennew, J. 1980. *The Western Isles today*. Cambridge: Cambridge University Press.

Ennis, P. 1962. The contextual dimension in voting. In *Public opinion and congressional elections*, W. McPhee & W. Glaser (eds.). New York: Free Press.

Epstein, L. K. 1979. Individual and contextual effects on partisanship. *Soc. Sci. Q.* **60**, 314–32.

Eulau, H. 1976. Understanding political life in America: the contribution of political science. *Soc. Sci. Q.* **57**, 112–53.

Evans, D. 1984. Demystifying suburban landscapes. In *Geography and the urban environment*, volume VI. D. T. Herbert & R. J. Johnston (eds.). Chichester: Wiley.

Eyles, J. 1985. *Senses of place*. Warrington: Silverbrook.

Faccioli, M. 1984. Geografia e spazio 'quotidiano.' *Riv. Geogr. Ital.* **91**, 215–38.

Faris, R. E. L. 1967. *Chicago sociology 1920–1932*. San Francisco: Chandler.

Farley, R. 1978. Chocolate city, vanilla suburbs: will the trend toward racially separate communities continue? *Soc. Sci. Q.* **59**, 319–44.

Fauvet, J. & H. Mendras 1958. *Les paysans et la politique dans la France contemporaine*. Paris: Armand Colin, Cah. Fond. Nat. Sci. Pol. **98**.

Fay, B. 1978. Practical reasoning, rationality and the explanation of intentional action. *J. Theory Soc. Behav.* **8**, 77–102.

Febvre, L. 1922. *La Terre et l'evolution humaine: introduction geographique à l'histoire*. Paris: La Renaissance du Livre, reprint edition 1949.

Fenno, R. F. 1978. *Home style*. Boston: Little, Brown.

Fenton, J. H. 1962. Ohio's unpredictable voters. *Harper's Mag.* **225**, 61–5.

Ferrarotti, F. 1981. On the autonomy of the biographical method. In *Biography and society: the life-history approach in the social sciences*, D. Bertaux (ed.). Beverly Hills, Calif.: Sage.

Ferrero, G. 1981. *Potere*. Milan: Sugarco.

Fieleke, N. S. 1981. Challenge and response in the automobile industry. *New Engl. Econ. Rev.* July–August, 37–48.

Firn, J. R. 1975. External control and regional development: the case of Scotland. *Env. Plann.* **A 7**, 393–414.

Fischer, C. S. 1975. The study of urban community and personality. *An. Rev. Soc.* 67–86.

Fitzpatrick, D. 1978. The geography of Irish Nationalism 1910–1921. *Past and Present* **78**, 113–44.

Flanagan, S. C. & R. J. Dalton 1984. Parties under stress: realignment and dealignment in advanced industrial societies. *W. Europ. Pol.* **7**, 7–23.

Flinn, T. A. 1962. Continuity and change in Ohio politics. *J. Pol.* **24**, 521–44.

Fogarty, M. 1945. *Prospects of the industrial areas of Great Britain*. London: Methuen.

Foster, J. 1975. *Class struggle and the industrial revolution: early industrial capitalism in three English towns*. New York: St. Martin's Press.

Foucault, M. 1975. *Surveiller et punir*. Paris: Gallimard.

Foucault, M. 1976. *Histoire de la sexualité: 1. La volonté de savoir*. Paris: Gallimard.

Foucault, M. 1980. *Power/knowledge: selected interviews and other writings*. Brighton: Harvester.

Fox, E. W. 1971. *History in geographic perspective: the other France*. New York: Norton.

Fox-Genovese, E. 1976. *The origins of physiocracy*. Ithaca, N.Y.: Cornell University Press.

Francis, J. G. & C. Payne 1977. The use of logistic models in political science: the British elections, 1964–70. *Pol. Meth.* **12**, 233–70.

Franklin, M. N. & A. Mughan 1978. Class voting in Britain: decline or disappearance? *Am. Pol. Sci. Rev.* **72**, 523–34.

Freeman, T. W. 1961. *A hundred years of geography*. London: Duckworth.

Gallagher, T. 1981. Catholics in Scottish politics. *Bull. Scott. Pol.* **2**, 21–43.

Gallagher, T. 1984. Protestant militancy and the Scottish working class before 1914. *Rad. Scot.* **10**, 23–5.

Galli, G. & A. Prandi 1970. *Patterns of political participation in Italy*. New Haven, Conn.: Yale University Press.

Gallie, W. B. 1955–6. Essentially contested concepts. *Proc. Arist. Soc.* **56**, 167–98.

Gaventa, J. 1980. *Power and powerlessness: quiescence and rebellion in an Appalachian valley*. Urbana, Ill.: University of Illinois Press.

Geertz, C. 1963. The integrative revolution. In *Old societies and new states*, C. Geertz (ed.). Glencoe, Ill.: Free Press.

George, P. 1966. *Sociologie et géographie*. Paris: Presses Universitaires de la France.

Gibbins, R. 1982. *Regionalism: territorial politics in Canada and the United States*. Toronto: Butterworths.

Giddens, A. 1979. *Central problems in social theory: action, structure and contradiction in social analysis*. London: Macmillan.

Giddens, A. 1981. *A contemporary critique of historical materialism*. Berkeley, Calif.: University of California Press.

Giddens, A. 1983. Comments on the theory of structuration. *J. Theory Soc. Behav.* **13**, 75–80.

Gillanders, F. 1968. The economic life of Gaelic Scotland today. In *The future of the highlands*, D. S. Thomson & I. Grimble (eds.). London: Routledge & Kegan Paul.

Glantz, O. 1958. Class-consciousness and political solidarity. *Am. Soc. Rev.* **23**, 375–83.

Glazer, S. 1965. *Detroit: a study in urban development*. New York: Bookman Associates.

Godelier, M. 1972. Structure and contradiction in capital. In *Ideology in social science*, Robin Blackburn (ed.). London: Fontana.

Goffman, E. 1972. The neglected situation. In *Language and social context*, P. P. Giglioli (ed.). London: Penguin.

Goguel, F. 1951a. Esquisse d'un bilan. In *Sociologie électorale: esquisse d'un bilan, guide de recherches*. Paris: Armand Colin, Cah. Fond. Nat. Sci. Pol. 26.

Goguel, F. 1951b. *Géographie des élections françaises de 1870 à 1951*. Paris: Armand Colin, Cah. Fond. Nat. Sci. Pol. 27.

Goguel, F. 1954. *Nouvelle études de sociologie électorale*. Paris: Armand Colin, Cah. Fond. Nat. Sci. Pol. 60.

Goguel, F. (ed.) 1962. *Le referendum du 8 janvier 1961*. Paris: Armand Colin, Cah. Fond. Nat. Sci. Pol. 119.

Goguel, F. 1970. *Geographie des élections françaises sous la IIIe et la IVe Republique*. Paris: Armand Colin.

Goguel, F. 1983. *Chroniques électorales . . . la cinquième république après de Gaulle*. Paris: Presses de la Fondation Nationale des Sciences Politiques.

Gottmann, J. (ed.) 1980. *Center and periphery: spatial variation in politics*. Beverly Hills, Calif.: Sage.

Goubert, P. 1971. Local history. *Daedalus* **100**, 113–27.

Gourevitch, P. 1979. The reemergence of 'peripheral nationalism': some comparative speculations on the spatial distribution of political leadership and economic growth. *Comp. Studs. Soc. Hist.* **21**, 303–22.

Gramsci, A. 1971. *Selections from the prison notebooks*. New York: International Publishers.

Granata, I. 1977. Il sindicalismo fascista all conquista di Milano "rossa" (gennaio–luglio 1922). *Il Risorgimento* **1–2**, 58–84.

Granata, I. 1980. Storia nazionale e storia locale: alcune considerazioni sulla problematica del fascismo delle origini (1919–1922). *Storia contemporanea* **11**, 503–44.

Grasmuck, S. 1980. Ideology of ethnoregionalism: the case of Scotland. *Pol. and Soc.* **9**, 471–94.

Green, L. 1982. Rational nationalists. *Pol. Studs* **30**, 236–46.

Gregory, D. 1978. *Ideology, science and human geography*. London: Hutchinson.

Gregory, D. 1982. *Regional transformation and industrial revolution: a geography of the Yorkshire woollen industry*. Minneapolis: University of Minnesota Press.

Groop. R. E. 1984. *Michigan political atlas*. East Lansing, Mich.: Michigan State University Center for Cartographic Research and Spatial Analysis.

Gurvitch, G. 1955. Le concept de la structure sociale. *Cah. intern. sociol.* **19**, 3–44.

Gurvitch, G. 1963. *Traité de sociologie*, 2 vols. Paris: Colin.

Gusfield, J. R. 1967. Tradition and modernity: misplaced polarities in the study of social change. *Am. J. Soc.* **72**, 351–62.

Gusfield, J. R. 1975. *Community: a critical response*. New York: Harper & Row.

Hache, J. D. 1982. Insularité et institutionalization dans les Hébrides-exterieures d'Écosse. *Rev. française sci. pol.* **32**, 743–67.

Hägerstrand, T. 1970. What about people in regional science? *Paps Reg. Sci. Assoc.* **24**, 7–21.

Haggett, P. 1965. *Locational analysis in human geography*. London: Edward Arnold.

Haider, D. H. 1974. *When governments come to Washington: governors, mayors, and intergovernmental lobbying*. New York: Free Press.

Hamilton, R. F. 1982. *Who voted for Hitler?* Princeton N.J.: Princeton University Press.

Hanby, V. J. 1977. Current Scottish nationalism. *Scot. J. Soc.* **1**, 95–109.

Handley, J. 1964. *The Irish in modern Scotland*. Cork: Cork University Press.

Hanham, H. J. 1959. *Elections and party management: politics in the time of Gladstone and Disraeli*. Reprinted 1978. Brighton: Harvester Press.

Hanham, H. J. 1969a. The problem of Highland discontent. *Trans. R. Hist. Soc.* 5th series, **19**, 21–65.

Hanham, H. J. 1969b. *Scottish nationalism*. Cambridge, Mass.: Harvard University Press.

Hannan, M. T. 1979. The dynamics of ethnic boundaries in modern states. In *National development and the world-system*, M. T. Hannan & J. Meyer (eds.). Chicago: University of Chicago Press.

Hansen, T. 1981. Transforming needs into expenditure decisions. In *Urban political economy*, K. Newton (ed.). London: Frances Pinter.

Harré, R. 1970. *The principles of scientific thinking*. London: Macmillan.

Harré, R. 1979. *Social being*. Oxford: Blackwell.

Harré, R. 1981. Philosophical aspects of the micro-macro problem. In *Advances in social theory and methodology: toward an integration of micro- and macro-sociologies*, K. Knorr-Cetina & A. V. Cicourel (eds.). Boston: Routledge & Kegan Paul.

Harré, R. & E. G. Madden 1975. *Causal powers*. Oxford: Blackwell.

Harris, E. E. 1957. Political power. *Ethics* **48**, 1–10.

Harris, R. 1984a. Residential segregation and class formation in the capitalist city. *Prog. Hum. Geogr.* **8**, 26–49.

Harris, R. 1984b. A political chameleon: class segregation in Kingston, Ontario, 1961–1976. *Ann. Assoc. Am. Geogs.* **74**, 454–76.

Hartz, L. 1955. *The liberal tradition in America: an interpretation of American political thought since the Revolution*. New York: Harcourt Brace.

Harvey, D. W. 1973. *Social justice and the city*. Baltimore: Johns Hopkins University Press.

Harvey, D. W. 1982. *The limits to capital*. Chicago: University of Chicago Press.

Harvie, C. 1981. *No gods and precious few heroes: Scotland, 1914–1980*. Toronto: University of Toronto Press.

Hauser, R. M. 1970. Context and consex: a cautionary tale. *Am. J. Soc.* **73**, 645–64.

Hays, S. P. 1967. Political parties and the community–society continuum. In *The American party systems: stages of political development*, W. N. Chambers & W. D. Burnham (eds.). New York: Oxford University Press.

Hechter, M. 1975. *Internal colonialism: the Celtic fringe in British national development*. Berkeley, Calif.: University of California Press.

Hechter, M. 1983. Introduction. In *The microfoundations of macrosociology*, M. Hechter (ed.). Philadelphia: Temple University Press.

Henderson, R. A. 1974. Industrial overspill from Glasgow: 1958–1968. *Urb. Studs.* **11**, 61–79.

Hill, R. C. 1984. Economic crisis and political response in motor city. In *Sunbelt/Snowbelt*, L. Sawers & W. K. Tabb (eds.). New York: Oxford University Press.

Hirschman, A. O. 1970. *Exit, voice and loyalty*. Cambridge, Mass.: Harvard University Press.

Hobsbawm, E. & T. Ranger (eds.) 1983. *The invention of tradition*. Cambridge: Cambridge University Press.

Hofferbert, R. I. 1971. Socioeconomic dimensions in the American states. In *State and urban politics*, R. I. Hofferbert & I. Sharkansky (eds.). Boston: Little, Brown.

Hofferbert, R. I. & I. Sharkansky 1971. The nationalization of state politics. In *State and urban politics*, R. I. Hofferbert & I. Sharkansky (eds.). Boston: Little, Brown.

Hoffman, S. 1956. *Le mouvement Poujade*. Paris: Armand Colin, Cah. Fond. Nat. Sci. Pol., 81.

Hoggart, K. 1986. Geography, political control and local government policy outputs. *Prog. Hum. Geog.* **10**, 1–23.

Hollingsworth, T. H. 1970. *Migration: a study based on Scottish experience between 1939 and 1964*. Edinburgh: Oliver & Boyd.

Hollis, M. 1977. *Models of man*. Cambridge: Cambridge University Press.

Hood, N. & S. Young 1982. *Multinationals in retreat: the Scottish experience*. New York: Columbia University Press.

Huckfeldt, R. R. 1979. Political participation and the neighborhood social context. *Am. J. Pol. Sci.* **23**, 579–92.

Huckfeldt, R. R. 1983. The social context of political change: durability, volatility, and social influence. *Am. Pol. Sci. Rev.* **77**, 929–44.

Hummon, D. M. 1986. City mouse, country mouse: the persistence of community identity. *Qualit. Soc.* **9**, 3–25.

Hunt, L. 1984. The political geography of revolutionary France. *J. Interdisc. Hist.* **14**, 535–59.

Hunter, J. 1974. The politics of highland land reform, 1873–1895. *Scot. Hist. Rev.* **53**, 45–68.

Hunter, J. 1975. The gaelic connection: the highlands, Ireland and nationalism. *Scot. Hist. Rev.* **54**, 52–71.

Hunter, J. 1976. *The making of the crofting community*. Edinburgh: Donald.

Huntington, S. P. 1968. *Political order in changing societies*. New Haven, Conn.: Yale University Press.

Irish, M. D. & J. W. Prothro 1968. *The politics of American democracy*, 4th edn. Englewood Cliffs, N.J.: Prentice-Hall.

Jackman, R. W. 1979. Political parties, voting, and national integration. In *The Canadian political process*, R. Schultz, O. M. Kruhlak & J. C. Terry (eds.). Toronto: Holt, Rinehart & Winston.

Jacoby, L. R. 1972. *Perception of air, noise, and water pollution in Detroit*. Ann Arbor, Mich.: University of Michigan Department of Geography.

Jaffré, J. 1980. The French electorate in March 1978. In *The French national assembly elections of 1978*, H. R. Penniman (ed.). Washington, D.C.: American Enterprise Institute for Public Policy Research.

James, P. E. & G. J. Martin 1981. *All possible worlds: a history of geographical ideas*. New York: Wiley.

Jensen, R. 1969a. History and the political scientist. In *Politics and the social sciences*, S. M. Lipset (ed.). New York: Oxford University Press.

Jensen, R. 1969b. American election analysis: a case history of methodological innovation and diffusion. In *Politics and the social sciences*, S. M. Lipset (ed.). New York: Oxford University Press.

Johnston, R. 1980. Federal and provincial voting: contemporary patterns and historical evolution. In *Small worlds: provinces and parties in Canadian political life*, D. J. Elkins & R. Simeon (eds.). Toronto: Methuen.

Johnston, R. J. 1976. Political behavior and the residential mosaic. In *Social areas in cities*, vol. II, D. Herbert & R. J. Johnston (eds.). London: Wiley.

Johnston, R. J. 1979. *Political, electoral and spatial systems*. London: Oxford University Press.

Johnston, R. J. 1981. Regional variations in British voting trends – 1966–1979: tests of an ecological model. *Reg. Studs.* **15**, 23–32.

Johnston,, R. J. 1982. The definition of voting regions in multi-party contests. *Eur. J. Pol. Res.* **10**, 293–304.

Johnston, R. J. 1983. Class locations, consumption locations and the geography of voting in England. *Soc. Sci. Res.* **12**, 215–35.

Johnston, R. J. 1985a. *The geography of English politics: voting at the 1983 election*. London: Croom Helm.

Johnston, R. J. 1985b. Class and the geography of voting in England: towards measurement and understanding. *Trans. Inst. Br. Geogs* **N.S. 10**, 245–55.

Johnston, R. J. & A. M. Hay 1982. On the parameters of uniform swing in single-member constituency electoral systems. *Env. Plann.* **A 14**, 61–74.

Johnston, R. J., A. B. O'Neill & P. J. Taylor 1985. *The geography of party support: comparative studies in electoral stability*. Newcastle-upon-Tyne: University of Newcastle, Department of Geography, Seminar Paper No. 40.

Johnston, T. L., N. K. Buxton & D. Mair 1971. *Structure and growth of the Scottish economy*. London: Collins.

Jones, G. S. 1983. Introduction. In *Languages of class: studies in English working class history, 1832–1982*. Cambridge: Cambridge University Press.

Joyce, P. 1975. The factory politics of Lancashire in the later nineteenth century. *Hist. J.* **18**, 525–53.

Joyce, P. 1980. *Work, society and politics: the culture of the factory in late Victorian England*. Hassocks, England: Harvester.

Judt, T. 1975. The development of socialism in France: the example of the Var. *Hist. J.* **19**, 55–83.

Judt, T. 1979. *Socialism in Provence, 1871–1914: a study in the origins of the modern French left*. Cambridge: Cambridge University Press.

Kasperson, R. & J. Minghi (eds.), 1969. *The structure of political geography*. Chicago: Aldine.

Katz, R. S. 1973. The attribution of variance in electoral returns: an alternative measurement technique. *Am. Pol. Sci. Rev.* **67**, 817–28.

Katznelson, I. 1976. Class capacity and social cohesion in American cities. In *The new urban politics*, L. H. Masotti & R. L. Lineberry (eds.). Cambridge, Mass.: Ballinger.

Katznelson, I. 1981. *City trenches: urban politics and the patterning of class in the United States*. New York: Pantheon.

Kauppi, M. 1982. The decline of the Scottish National Party, 1977–81: political and organizational factors. *Ethn. Rac. Studs* **5**, 326–48.

Keat, R. 1971. Positivism, naturalism, and anti-naturalism in the social sciences. *J. Theory Soc. Behav.* **1**, 3–17.

Keat, R. & J. Urry 1975. *Social theory as science*. London: Routledge & Kegan Paul.

Kellas, J. G. 1979. *The Scottish political system*. Cambridge: Cambridge University Press.

Kellas, J. G. & P. Fotheringham 1976. The political behaviour of the working class. In *Social class in Scotland: past and present*, A. A. MacLaren (ed.). Edinburgh: Donald.

Kershaw, I. 1983. *Popular opinion and political dissent in the Third Reich: Bavaria 1933–1945*. Oxford: Clarendon Press.

Key, V. O. 1949. *Southern politics in state and nation*. New York: Knopf.

Key, V. O. 1955. A theory of critical elections. *J. Pol.* **17**, 3–18.

Key, V. O. 1964. *Politics, parties and pressure groups*, 5th edn. New York: Crowell.

Kiewiet, D. R. 1983. *Macroeconomics and micropolitics*. Chicago: University of Chicago Press.

Kinnear, M. 1968. *The British voter: an atlas and survey since 1885*. Ithaca, N.Y.: Cornell University Press.

Kirby, J. T. 1984. The South as pernicious abstraction. *Perspect. Am. Sth* **2**, 167–80.

Klatzmann, J. 1957. Comportement électoral et classe sociale: étude du vote communiste et du vote socialiste a Paris et dans la Seine. In *Les elections du 2 janvier 1956*, M. Duverger *et al.* (eds.). Paris: Armand Colin, Cah. Fond. Nat. Sci. Pol. 82.

Kleppner, P. 1982. *Who voted? the dynamics of electoral turnout, 1870–1980*. New York: Praeger.

Klingemann, H. & F. Pappi 1970. The 1969 Bundestag election in the Federal Republic of Germany. *Comp. Pol.* **2**, 523–48.

Knorr-Cetina, K. 1981. The micro-sociological challenge of macro-sociology: towards a reconstruction of social theory and methodology. In *Advances in social theory and methodology: toward an integration of micro- and macro-sociologies*, K. Knorr-Cetina & A. Cicourel (eds.). Boston: Routledge & Kegan Paul.

Knorr-Cetina, K. & A. Cicourel (eds.) 1981. *Advances in social theory and methodology: toward an integration of micro- and macro- sociologies*. Boston: Routledge & Kegan Paul.

Knox, P. L. 1979. Territorial social indicators and area profiles: some cautionary observations. *Town Plann. Rev.* **49**, 75–83.

Knox, P. L. & M. B. Cottam 1981. Rural deprivation in Scotland: a preliminary assessment. *Tijd. Econ. Soc. Geogr.* **72**, 162–75.

Kornhauser, W. 1959. *The politics of mass society*. New York: Free Press.

Kyd, J. G. (eds.) 1952. *Scottish population statistics including Webster's analysis of population 1755*. Edinburgh: Scottish History Society.

Laing, R. D. 1956. *The divided self*. London: Tavistock.

Lancelot, A. 1968. *L'abstentionnisme électoral en France*. Paris: Armand Colin, Cah. Fond. Nat. Sci. Pol. 162.

Lanzillo, A. 1922. *Le rivoluzioni del dopoguerra: critica e diagnosi*. Città di Castello: Lanzillo.

Laslett, B. 1980. The place of theory in quantitative historical research. *Am. Soc. Rev.* **45**, 214–28.

Laslett, P. 1956. The face to face society. In *Philosophy, politics and society*, P. Laslett (ed.). Oxford: Basil Blackwell.

Lawson, S. F. 1976. *Black ballots: voting rights in the South, 1944–1967*. New York: Columbia University Press.

Layder, D. 1985. Power, structure and agency. *J. Theory Soc. Behav.* **15**, 131–50.

Lazarsfeld, P. F., B. R. Berelson & H. Gaudet 1948. *The People's Choice*. New York: Columbia University Press.

LeBras, G. 1949. Géographie politique et géographie religieuse. In *Études de sociologie électorale*, C. Morazé et al. (eds.). Paris: Armand Colin, Cah. Fond. Nat. Sci. Pol. 1.

LeBras, H. & E. Todd 1981. *L'Invention de la France: atlas anthropologique et politique*. Paris: Librairie Générale Française.

Leca, J. 1982. La science politique dans le champ intellectuel français, *Rev. française sci. pol.* **32**, 653–78.

Lee, C. H. 1983. Modern economic growth and structural change in Scotland: the service sector reconsidered. *Scot. Econ. Soc. Hist.* **3**, 5–35.

Lefebvre, H. 1971. *Everyday life in the modern world*. London: Allen Lane.

Lefebvre, H. 1976. *The survival of capitalism*. London: Allison & Busby.

LeLannou, M. 1977. *Europe, terre promise*. Paris: Seuil.

Lenman, B. 1981. *Integration, enlightenment and industrialization: Scotland 1746–1832*. Toronto: University of Toronto Press.

Lerner, D. 1958. *The passing of traditional society*. Glencoe, Ill.: Free Press.

Lernoux, P. 1984. The Miami Connection. *Nation*. Feb. 18, 186–98.

Letwin, W. 1963. *The origins of scientific economics: English economic thought 1660–1776*. London: Methuen.

Leuilliot, P. 1974. Histoire locale et politique de l'histoire. Preface to *Economie et société nivernaise au debut du XIXe siècle*, by Guy Thuillier. Paris: Mouton.

Lévy, J. 1984a. Paris, carte d'identité: espace géographique et sociologie politique. In *Sens et non-sens de l'espace*. Paris: Collectif français de géographie urbaine et sociale.

Lévy, J. 1984b. Espace et politique: une nouvelle chance. *Espaces–Temps*, **26–8**, 91–9.

Ley, D. 1978. Social geography and social action. In *Humanistic geography: prospects and problems*, D. Ley & M. S. Samuels (eds.). Chicago: Maaroufa.

Ley, D. 1984. Pluralism and the Canadian state. In *Geography and ethnic pluralism*, C. Clarke, D. Ley & C. Peach (eds.). London: Allen & Unwin.

Ley, D. & M. S. Samuels (eds.) 1978. *Humanistic geography: prospects and problems*. Chicago: Maaroufa.

Lipset, S. M. 1960. Economic development and democracy. In S. M. Lipset, *Political man*. London: Heinemann.

Lipset, S. M. 1963. *The first new nation: the United States in historical and comparative perspective*. New York: Doubleday.

Lipset, S. M. 1981. The revolt against modernity. In *Mobilization, center–periphery structures and nation-building*, P. Torsvik (ed.). Oslo: Universitetsforlaget.

Lipset, S. M. & S. Rokkan 1967. Cleavage structures, party systems, and voter alignments: an introduction. In *Party systems and voter alignments: cross-national perspectives*, S. M. Lipset & S. Rokkan (eds.). New York: Free Press.

Little, J. S. 1985. Foreign direct investment in New England. *New Engl. Econ. Rev.* March–April, 48–57.

Livingstone, D. N. 1984. Natural theology and neo-Lamarckism: the changing context of nineteenth-century geography in the United States and Great Britain. *Ann. Assoc. Am. Geogs.* **74**, 9–28.

Lockwood, D. 1966. Sources of variation in working-class images of society. *Soc. Rev.* **14**, 249–67.

Logan, J. R. 1976a. Industrialization and the stratification of cities in suburban regions. *Am. J. Soc.* **82**, 333–48.

Logan, J. R. 1976b. Notes on the growth machine – toward a *comparative* political economy of place. *Am. J. Soc.* **82**, 349–51.

Logan, J. R. 1978. Growth, politics and the stratification of places. *Am. J. Soc.* **84**, 404–16.

Louch, A. R. 1966. *Explanation and human action.* Oxford: Basil Blackwell.

Luger, M. I. 1984. Federal tax incentives as industrial and urban policy. In *Sunbelt/snowbelt*, L. Sawers & W. K. Tabb (eds.). New York: Oxford University Press.

Lukas, J. A. 1984. Our town 1984: the election comes to Mansfield, Ohio. *New York Times Mag.* Oct. 21, 26–40 *et seq.*

Lukes, S. 1973. *Individualism.* Oxford: Blackwell.

Lukes, S. 1975. *Power: a radical view.* London: Macmillan.

Lukes, S. 1977. *Essays in social theory.* London: Macmillan.

Luria, D. & J. Russell 1984. Motor city changeover. In *Sunbelt/snowbelt*, L. Sawers & W. K. Tabb (eds.). New York: Oxford University Press.

Lythe, C. & M. Majmudar 1982. *The renaissance of the Scottish economy.* London: Allen & Unwin.

MacKenzie, B. D. 1977. *Behaviourism and the limits of scientific method.* Atlantic Highlands, N.J.: Humanities Press.

Mackenzie, W. J. M. 1981. Peripheries and nation-building: the case of Scotland. In *Mobilization, center–periphery structures and nation-building*, P. Torsvik (ed.). Oslo: Universitetsforlaget.

MacLaughlin, J. 1987. The political geography of nationalism and nation-building in social science: structural versus dialectical accounts. *Pol. Geog. Q.* **6**, in press.

MacLaughlin, J. G. & J. A. Agnew 1986. *Hegemony* and the regional question: the political geography of regional industrial policy in Northern Ireland, 1945–72. *Ann. Assoc. Am. Geogs.* **76**, 247–61.

Macpherson, C. B. 1962. *The political theory of possessive individualism: Hobbes to Locke.* New York: Oxford University Press.

Madgwick, P. & R. Rose (eds.) 1982. *The territorial dimension in United Kingdom politics.* London: Macmillan.

Mair, R. & I. McAllister 1982. A territorial *versus* a class appeal? the Labour parties of the British Isles' periphery. *Eur. J. Pol. Res.* **10**, 17–34.

Mann, T. E. 1978. *Unsafe at any margin.* Washington, D.C.: American Enterprise Institute for Public Policy Research.

Mannheimer, R. 1980. Un'analisi territoriale del voto communista. In *Mobilità senza movimento: le elezioni del 3 giugno 1979*, A. Parisi (ed.). Bologna: Mulino.

Margadant, T. W. 1981. Proto-urban development and political mobilization during the second republic. In *French cities in the nineteenth century*, J. M. Merriman (ed.). New York: Holmes & Meier.

Marie, C. 1965. *L'evolution du comportement politique dans une ville en expansion, Grenoble, 1871–1965.* Paris: Armand Colin, Cah. Fond. Nat. Sci. Pol. 48.

Massey, D. 1984. *Spatial divisions of labor: social structures and the geography of production.* London: Methuen.

McAllister, I. 1981. Party organization and minority nationalism: a comparative study in the United Kingdom. *Eur. J. Pol. Res.* **9**, 237–55.

McAllister, I. 1983. Territorial differentiation and party development in Northern Ireland. In *Contemporary Irish studies*, T. Gallagher & J. O'Connell (eds.). Manchester: Manchester University Press.

McAllister, I. 1987. Social context, turnout, and the vote: Australian and British comparisons. *Pol. Geog. Q.* **6**, 17–30.

McCaffrey, J. 1970. The Irish vote in Glasgow in the later nineteenth century: a preliminary survey. *Innes Rev.* **21**, 30–7.

McDonald, G. 1975. Social and geographical mobility in the Scottish new towns. *Scot. Geog. Mag.* **91**, 38–51.

McKinney, J. C. & L. B. Bourque 1971. The changing South: national incorporation of a region. *Am. Soc. Rev.* **36**, 399–412.

McLean, I. 1973. The problem of proportionate swing. *Pol. Studs* **21**, 57–63.

McLean, I. 1977. The politics of nationalism and devolution. *Pol. Studs* **25**, 425–30.

McLean, I. 1983. *The legend of red Clydeside*. Edinburgh: Donald.

McWilliams, W. C. 1972. The American constitutions. In *The performance of American government*, G. Pomper *et. al.* (eds.). New York: Free Press.

Meek, R. 1976. *Turgot on progress, sociology, and economics*. Cambridge: Cambridge University Press.

Menzel, H. 1950. Comment on Robinson's "ecological correlations and the behavior of individuals." *Am. Soc. Rev.* **15**, 674.

Merriman, J. M. 1979. Incident at the statue of the Virgin Mary: the conflict of old and new in nineteenth-century Limoges. In *Consciousness and class experience in nineteenth-century Europe*, J. M. Merriman (ed.). New York: Holmes & Meier.

Merriman, J. M. 1981. Restoration town, bourgeois city: changing urban politics in industrializing Limoges. In *French cities in the nineteenth century*, J. M. Merriman (ed.). New York: Holmes & Meier.

Merton, R. 1968. *Social theory and social structure*. New York: Free Press.

Meyrowitz, J. 1985. *No sense of place: the impact of electronic media on social behavior*. New York: Oxford University Press.

Miliband, R. 1969. *The state in capitalist society*. London: Weidenfeld & Nicolson.

Miliband, R. 1977. *Marxism and politics*. London: Oxford University Press.

Miller, W. E. & D. E. Stokes 1966. Party government and the salience of Congress. In *Elections and the political order*, A. Campbell, P. E. Converse, W. E. Miller & D. E. Stokes (eds.). New York: Wiley.

Miller, W. L. 1979. Class, region, and strata at the British general election of 1979. *Parl. Aff.* **32**, 376–82.

Miller, W. L. 1981. *The end of British politics? Scots and English political behaviour in the seventies*. Oxford: Clarendon Press.

Miller, W. L. 1984. The de-nationalisation of British politics: the re-emergence of the periphery. In *Change in British politics*, H. Berrington (ed.). London: Frank Cass.

Miller, W. L. & G. Raab 1977. The religious alignment at English elections between 1918 and 1970. *Pol. Studs.* **25**, 227–51.

Miller, W. L., B. Särlvik, I. Crewe & J. Alt 1977. The connection between SNP voting and demand for Scottish self-government. *Eur. J. Pol. Res.* **5**, 88–102.

Mitchell, B. R. & K. Boehm 1966. *British parliamentary election results, 1850–1964*. Cambridge: Cambridge University Press.

Mollenkopf, J. H. 1983. *The contested city*. Princeton, N.J.: Princeton University Press.

Molotch, H. 1976. The city as a growth machine: toward a political economy of place. *Am. J. Soc.* **82**, 309–32.

Moore, B. 1966. *Social origins of dictatorship and democracy*. Boston: Beacon Press.

Moore, R. 1982. *The social impact of oil: the case of Peterhead*. London: Routledge & Kegan Paul.

Morazé, C. 1949. Quelques questions de methode. In *Études de sociologie électorale*, C. Morazé et. al. (eds.). Paris: Armand Colin, Cah. Fond. Nat. Sci. Pol. 1.

Mosse, G. L. 1975. *The nationalization of the masses: political symbolism and mass movements in Germany from the Napoleonic Wars through the Third Reich*. New York: Howard Fertig.

Muir, R. & R. Paddison 1981. *Politics, geography and behaviour*. London: Methuen.

Muscarà, C. 1983. Culture locali tra geografia e ideologia. 23rd Congress of Italian Geography, Catania.

Nairn, T. 1977. *The break-up of Britain*. London: New Left Books.

Naughtie, J. 1976. Grampian seats. *Q*, May 26.

Nettl, J. P. 1967. *Political mobilization: a sociological analysis of methods and concepts*. London: Faber

Newton, K. 1978. Conflict avoidance and conflict suppression: the case of urban politics in the United States. In *Urbanization and conflict in market societies*, K. R. Cox (ed.). Chicago: Maaroufa.

Nielsen, F. 1980. The Flemish movement in Belgium after World War II: a dynamic analysis. *Am. Soc. Rev.* **45**, 76–94.

Niemi, R. & H. Weisberg (eds.) 1976. *Controversies in American voting behavior*. New York: W. H. Freeman.

Nisbet, R. A. 1953. *The quest for community*. New York: Oxford University Press.

Nisbet, R. A. 1966. *The sociological tradition*. London: Heinemann.

Nordheimer, J. 1985a. Cuban-Americans in Miami move to threshold of power. *New York Times* Nov. 7, A1, B11.

Nordheimer, J. 1985b. Miami's first Cuban-born mayor. *New York Times* Nov. 14, A24.

Nordheimer, J. 1985c. Miami cultures: collision then combination. *New York Times* Dec. 16, A1, A16.

Nossiter, T. 1970. Voting behavior 1832–1872. *Pol. Studs.* **18**, 380–9.

O'Neill, J. 1973. *Modes of individualism and collectivism*. London: Heinemann.

Orridge, A. W. 1981. Uneven development and nationalism: 1, 2. *Pol. Studs* **29**, 1–15, 181–90.

Orridge, A. W. & C. H. Williams 1982. Autonomist nationalism: a theoretical framework for spatial variations in its genesis and development. *Pol. Geog. Q.* **1**, 19–39.

Osmond, J. 1977. *Creative conflict: the politics of Welsh devolution*. London: Routledge & Kegan Paul.

Owsley, F. L. 1949. *Plain folk of the old South*. Baton Rouge: Louisiana State University Press.

Pacione, M. 1972. Traditional and new industries in Dundee. *Scot. Geog. Mag.* **88**, 53–60.

Paige, J. 1975. *Agrarian revolution*. New York: Free Press.

Parisi, A. (ed.) 1984. *Luoghi e misure della politica*. Bologna: Mulino.

Parker, W. O. 1982. *Mackinder: geography as an aid to statecraft*. Oxford: Clarendon Press.

Parodi, J. L. 1978. L'échec des gauches. *Rev. pol. parl.* **873**, 9–32.

Parsons, T. 1937. *The structure of social action*, New York: McGraw Hill.

Parsons, T. 1951. *The social system*. New York: Free Press.

Parsons, T. 1959. Voting and the equilibrium of the American political system. In *American voting behavior*, E. Burdick & A. Brodbeck (eds.). New York: Free Press.

Parsons, T. & E. A. Shils 1951. *Toward a general theory of action*. Cambridge, Mass.: Harvard University Press.

Parsons, T., R. F. Bales & E. A. Shils 1953. *Working papers in the theory of action*. Glencoe, Ill.: Free Press.

Passcher, M. 1980. The electoral geography of the Nazi landslide: the need for community studies. In *Who were the Fascists? Social roots of European Fascism*, S. U. Larsen, B. Hagtvet & J. P. Myklebust (eds.). Oslo: Universitetsforlaget.

Pavsič, R. 1985. Esiste una tendenza all' omogeneizzazione territoriale nei partiti Italiani? *Riv. Ital. Sci. Pol.* **15**, 70–97.

Pederson, M. 1983. Changing patterns of electoral volatility in European party systems. In *Western European party systems*, H. Daalder & P. Mair (eds.). Beverly Hills, Calif.: Sage.

Pelling, H. 1967. *Social geography of British elections, 1885–1910*. London: Macmillan.

Percheron, A. 1982. The influence of the socio-political context on political socialization. *Eur. J. Pol. Res.* **10**, 53–69.

Perin, C. 1977. *Everything in its place: social order and land use in America*. Princeton N.J.: Princeton University Press.

Perrot, J-C. 1984. The golden age of regional statistics (Year IV: 1804). In *State and statistics in France, 1789–1815*, J-C. Perrot & S. J. Woolf. Chur, Switzerland: Harwood.

Peterson, P. 1979. Redistributive policies and patterns of citizen participation in local politics in the U.S.A. In *Decentralist trends in Western democracies*, L. J. Sharpe (ed.). London: Sage.

Phillips, D. C. 1976. *Holistic thought in social science*. Stanford, Calif.: Stanford University Press.

Phillips, K. 1969. *The emerging Republican majority*. New York: Doubleday.

Phillips, K. 1978. The balkanization of America. *Harper's Mag.* **256**, 37–47.

Phillips, K. 1982. *Post-conservative America*. New York: Random House.

Pierce, R. 1980. French legislative elections: the historical background. In *The French National Assembly elections of 1978*, H. R. Penniman (ed.). Washington, D.C.: American Enterprise Institute for Public Policy Research.

Pletsch, C. E. 1981. The three worlds, or the division of social scientific labor, circa 1950–1975. *Comp. Studs. Soc. Hist.* **23**, 565–90.

Polsby, N. W. 1981. The Washington community, 1960–1980. In *The new Congress*, T. E. Mann & N. J. Ornstein (eds.). Washington, D.C.: American Enterprise Institute for Public Policy Research.

Polsby, N. W. & A. Wildavsky 1980. *Presidential elections: strategies of American electoral politics*, 5th edn. New York: Scribners.

Pomper, G. with S. Lederman 1980. *Elections in America*. New York: Longman.

Poole, K. T. & H. Rosenthal 1984. US presidential elections 1968–1980: a spatial analysis. *Am. J. Pol. Sci.* **28**, 282–312.

Poulantzas, N. 1973. *Political power and social classes*. London: New Left Books.

Pounds, N. J. G. & S. S. Ball 1964. Core areas and the development of the European states system. *Ann. Assoc. Am. Geogs.* **54**, 24–40.

Pred, A. 1983. Structuration and place: on the becoming of sense of place and structure of feeling. *J. Theory Soc. Behav.* **13**, 45–68.

Pred, A. 1984. Place as historically contingent process: structuration and the time–geography of becoming places. *Ann. Assoc. Am. Geogs.* **74** (2), 279–97.

Przeworski, A. 1974. Contextual models of political behavior. *Pol. Meth.* **1**, 27–61.

Pulzer, P. J. 1967. *Political representation and elections in Britain*. London: Allen & Unwin.

Putnam, R. D. 1966. Political attitudes and the local community. *Am. Pol. Sci. Rev.* **60**, 640–54.

Rae, D. W. & M. Taylor 1970. *The analysis of political cleavages*. New Haven, Conn.: Yale University Press.

Raffestin, C. 1981. *Pour une géographie du pouvoir*. Paris: Les Librairies Techniques.

Ragin, C. 1976. Review of Michael Hechter's "Internal colonialism." *Social Forces* **55**, 553–4.

Ranger, T. O. 1976. From humanism to the science of man: colonialism in Africa and the understanding of alien societies. *Trans. R. Hist. Soc.* **26**, 115–41.

Redfield, R. 1930. *Tepoztlan: a Mexican village*. Chicago: University of Chicago Press.

Redfield, R. 1955. *The little community*. Chicago: University of Chicago Press.

Regional trends 1981. London: Central Statistical Office.

Relph, E. 1976. *Place and placelessness*. London: Pion.

Report of the President's Commission for a National Agenda for the Eighties 1981. *A national agenda for the eighties*. New York: Mentor.

Rich, R. 1981. *Causes and effects of inequality in urban services*. Lexington, Mass.: Lexington.

Robertson, D. J. 1958. Population, past and present. In *The third statistical account of Scotland: Glasgow*, J. Cunnison & J. Gifillan (eds.). Glasgow: Collins.

Robinson, J. 1962. *Economic philosophy*. London: Penguin.

Robinson, W. S. 1950. Ecological correlations and the behavior of individuals, *Am. Soc. Rev.* **15**, 351–7.

Rochefort, R. 1983. Le Territoire de la vie quotidienne et le referential habitant. *Géopoint '82* (Avignon), 257–61.

Rokkan, S. 1970. *Citizens, elections, parties*. Oslo: Universitetsforlaget.

Rokkan, S. & D. W. Urwin (eds.) 1982. *The politics of territorial identity: studies in European regionalism*. Beverly Hills, Calif.: Sage.

Rokkan, S. & D. W. Urwin 1983. *Economy, territory, identity: politics of west European peripheries*. Beverly Hills, Calif.: Sage.

Root, E. 1916. *Addresses on government and citizenship*, R. Bacon & J. B. Scott (eds.). Cambridge, Mass.: Harvard University Press.

Rose, H. & D. Deskins 1980. Felony murder: the case of Detroit. *Urb. Geogr.* **1**, 1–21.

Rose, R. 1970. *The United Kingdom as a multinational state*. Glasgow: University of Strathclyde Occasional Paper No. 6.

Rose, R. 1971. Class and party divisions: Britain as a test case. In *European politics: a reader*, M. Dogan & R. Rose (eds.). London: Macmillan.

Rose, R. 1974. Britain: simple abstractions and complex realities. In *Electoral behaviour: a comparative handbook*, R. Rose (ed.). New York: Free Press.

Rose, R. & D. W. Urwin 1969. Social cohesion, political parties and strains in regimes. *Comp. Pol. Studs.* **2**, 7–67.

Rose, R. & D. W. Urwin 1970. Persistence and change in Western party systems since 1945. *Pol. Studs.* **18**, 287–319.

Rose, R. & D. W. Urwin 1975. *Regional differentiation and political unity in Western nations*. Beverly Hills, Calif.: Sage.

Rosenberg, S. W. 1985. Sociology, psychology, and the study of political behavior: the case of research on political socialization, *J. Pol.* **47**, 715–31.

Rossiter, C. 1960. *Parties and politics in America*. Ithaca, N.Y.: Cornell University Press.

Rostow, W. W. 1960. *The stages of economic growth: a non-communist manifesto*. New York: Cambridge University Press.

Rumpf, E. & A. Hepburn 1977. *Nationalism and socialism in twentieth-century Ireland*. New York: Barnes & Noble.

Runciman, W. G. 1969. *Social science and political theory*. Cambridge: Cambridge University Press.

Russell, T. P. 1970. The size structure of Scottish agriculture. *Scot. Agric. Econ.* **20**, 299–325.
Ryan, A. 1970. *The philosophy of the social sciences.* London: Macmillan.

Sack, R. D. 1973. A concept of physical space in geography. *Geogr. Anal.* **5**, 16–34.
Sack, R. D. 1974. The spatial separatist theme in geography. *Econ. Geog.* **50**, 1–19.
Sani, G. 1976. Political traditions as contextual variables: partisanship in Italy. *Am. Pol. Sci. J.* **20**, 365–406.
Sani, G. 1977. The Italian electorate in the mid-1970s: beyond tradition? In *Italy at the polls: the parliamentary elections of 1976*, H. Penniman (ed.). Washington, D.C.: American Enterprise Institute for Public Policy Research.
Särlvik, B. & I. Crewe 1983. *Decade of dealignment: the conservative victory of 1979 and electoral trends in the 1970s.* Cambridge: Cambridge University Press.
Sartre, J.-P. 1959, *L'existentialisme est un humanisme.* Paris: Nagel.
Sartre, J.-P. 1960. *Questions de methode.* Paris: Gallimard.
Saunders, P. 1979. *Urban politics: a sociological interpretation.* London: Hutchinson.
Saunders, P. 1985. Space, the city and urban sociology. In *Social relations and spatial structures*, D. Gregory & J. Urry (eds.). London: Macmillan.
Savage, D. C. 1961. Scottish politics, 1885–86, *Scott. Hist. Rev.* **130**, 118–35.
Sayer, A. 1984. *Method in social science: a realist approach.* London: Hutchinson.
Schattschneider, E. E. 1960. *The semisovereign people.* New York: Holt, Rinehart & Winston.
Schlesinger, J. A. 1985. The new American political party. *Am. Pol. Sci. Rev.* **79**, 1152–69.
Schutz, A. 1967. *The phenomenology of the social world.* Evanston, Ill.: Northwestern University Press.
Scivoletto, A. 1983. 'Filosofia' del territorio. In *Sociologia del territorio*, A. Scivoletto (ed.). Milan: Angeli.
Scott, J. & M. Hughes 1976. Ownership and control in a satellite economy: a discussion from Scottish data. *Soc.* **10**, 21–41.
Sharpe, L. J. 1982. The Labour party and the geography of inequality. In *The politics of the Labour party*, D. Kavanagh (ed.). London: Allen & Unwin.
Sharpe, L. J. & K. Newton 1984. *Does politics matter? The determinants of public policy.* Oxford: Clarendon Press.
Sheingold, C. 1973. Social networks and voting: the resurrection of a research agenda. *Am. Soc. Rev.* **38**, 712–20.
Shibutani, T. 1955. Reference groups as perspectives. *Am. J. Soc.,* **60**, 562–9.
Shils, E. 1970. The contemplation of society in America. In *Paths of American thought*, M. White & A. Schlesinger Jr. (eds.). Boston: Houghton Mifflin.
Shortridge, J. R. 1976. Patterns of religion in the United States. *Geog. Rev.* **66**, 420–34.
Siegfried, A. 1913. *Tableau politique de la France de l'ouest.* Paris: Armand Colin.
Simmel, G. 1971. *On individuality and social forms: selected writings.* Chicago: University of Chicago Press.
Skinner, Q. 1978. *The foundations of modern political thought*, Volume 2. Cambridge: Cambridge University Press.
Skocpol, T. 1979. *States and social revolutions.* Cambridge: Cambridge University Press.
Slaven, A. 1975. *The economic development of the West of Scotland, 1760–1860.* London: Macmillan.
Sloan, W. N. 1979. Ethnicity or imperialism? A review article. *Comp. Studs. Soc. Hist.* **21**, 113–25.
Smith, A. D. 1979. *Nationalism in the twentieth century.* Oxford: Martin Robertson.

Smith, B. H. & J. L. Rodriguez 1974. Comparative working-class political behavior. *Am. Behav. Sci.* **18**, 73–92.

Smith, C. 1984. Local history in global context: social and economic transitions in western Guatemala. *Comp. Studs. Soc. Hist.* **26**, 193–228.

Smith, J. 1984. Labour tradition in Glasgow and Liverpool. *Hist. Workshop* **17**, 32–56.

Smith, N. 1981. Degeneracy in theory and practice: spatial interactionism and radical eclecticism. *Prog. Hum. Geog.* **5**, 111–18.

Smith, N. 1984. *Uneven development.* Oxford: Blackwell.

Smith, S. J. 1984. Practicing humanistic geography. *Ann. Assoc. Am. Geogs.* **74**, 353–74.

Sofen, E. 1961. The politics of metropolitan leadership: the Miami experience. *Midwest Pol. Sci. Rev.* **5**, 18–38.

Sofen, E. 1963. *The Miami metropolitan experiment.* Bloomington, Ind.: Indiana University Press.

Sopher, D. E. 1973. Place and location: notes on the spatial patterning of culture. *Soc. Sci. Q.* **53**, 321–37.

Sorauf, F. J. 1980. *Party politics in America.* Boston: Little, Brown.

Stern, A. 1975. Political legitimacy in local politics – the Communist party in northeastern Italy. In *Communism in Italy and France*, D. L. M. Blackmer & S. Tarrow (eds.). Princeton, N.J.: Princeton University Press.

Stinchcombe, A. 1965. Social structure and organizations. In *Handbook of organizations*, J. G. March (ed.). New York: Rand McNally.

Stoianovich, T. 1976. *French historical method: the Annales paradigm.* Ithaca, N.Y.: Cornell University Press.

Stokes, D. E. 1965. A variance components model of political effects. In *Mathematical applications in political science*, J. M. Claunch (ed.). Dallas: The Arnold Foundation.

Stokes, D. E. 1967. Parties and the nationalization of electoral forces. In *The American party systems: stages of political development*, W. N. Chambers & W. D. Burnham (eds.). New York: Oxford University Press.

Stone, N. O. 1982. Pillars of the Third Reich. *New York Rev. Books* **29**, May 13, 24–6.

Strassoldo, R. 1983. La sociologia e le scienze del territorio. In *Sociologia del territorio*, A. Scivoletto (ed.). Milan: Angeli.

Sundquist, J. L. 1973. *Dynamics of the party system: alignment and realignment of political parties in the United States.* Washington, D.C.: The Brookings Institution.

Suppa, S. 1984. Antonio Gramsci: americanismo e fordismo. In *Parsons politico: una lettura dell' americanismo.* Bari: Daedalo.

Suttles, G. D. 1972. *The social construction of communities.* Chicago: University of Chicago Press.

Sztompka, P. 1982. *Sociological dilemmas: toward a dialectic paradigm.* New York: Academic Press.

Szymanski, R. & J. A. Agnew 1981. *Order and skepticism: human geography and the dialectic of science.* Washington, D.C.: Association of American Geographers, Resource Publications.

Tarrow, S. 1977. *Between center and periphery: grassroots politicians in Italy and France.* New Haven, Conn.: Yale University Press.

Tabboni, S. 1985. Il tempo e lo spazio nella teoria sociologia. *Rass. Ital. Soc.* **26**, 569–96.

Tate, C. N. 1974. Individual and contextual variables in British voting behavior: an exploratory note. *Am. Pol. Sci. Rev.* **68**, 1656–62.

Taylor, A. H. 1973. The electoral geography of Welsh and Scottish nationalism. *Scot. Geog. Mag.* **93**, 44–52.

Taylor, C. 1971. Interpretation and the science of man. *Rev. Metaphys.* **25**, 1–32, 35–45.

Taylor, P. J. & R. J. Johnston 1979. *Geography of elections.* London: Penguin.

Thompson, E. P. 1972. *The making of the English working class.* London: Penguin.

Thompson, E. P. 1978. *The poverty of theory and other essays.* London: Merlin Press.

Thrift, N. J. 1983. On the determination of social action in space and time. *Soc. and Space* **1**, 23–57.

Thuillier, G. 1974. *Economie et société nivernaise au début du XIXᵉ siècle.* Paris: Mouton.

Tilly, C. (ed.) 1975. *The formation of national states in Western Europe.* Princeton, N.J.: Princeton University Press.

Tilly, C. 1973. Do communities act? *Soc. Inq.* **43**, 209–40.

Tilly, C. 1978. *From mobilization to revolution.* Reading, Mass.: Addison-Wesley.

Tilly, C. 1979. Did the cake of custom break? In *Consciousness and class experience in nineteenth-century Europe,* J. M. Merriman (ed.). New York: Holmes & Meier.

Tingsten, H. 1937. *Political behavior: studies in election statistics,* reprinted 1963. Totowa, N.J.: Bedminster Press.

Tipps, D. 1973. Modernization theory and the comparative study of societies: a critical perspective. *Comp. Studs. Soc. Hist.* **15**, 199–226.

de Tocqueville, A. 1945. *Democracy in America,* revised edition. New York: Knopf.

Tönnies, F. 1887. *Community and association.* London: Routledge & Kegan Paul.

Trachte, K. & R. Ross 1985. The crisis of Detroit and the emergence of global capitalism. *Int. J. Urb. Reg. Res.* **9**, 186–217.

Trigilia, C. 1981. Struttura di classe e sistema politico in Italia: verso un modello neolocalistico? In *I ceti medi in Italia,* C. Carboni (ed.). Bari: Laterza.

Tuan, Y-F. 1974. Space and place: humanistic perspective. *Prog. Geog.* **6**, 211–52.

Tufte, E. 1973. The relationship between votes and seats in two-party systems. *Am. Pol. Sci. Rev.* **68**, 540–54.

Turner, B. S. 1981. Marginal politics, cultural identities and the clergy in Scotland. *Int. J. Soc. Soc. Pol.* **1**, 89–113.

Turner, F. J. 1932. *Sections in American history.* New York: Holt.

Turner, J. 1970. *Party and constituency: pressures on Congress.* Baltimore: Johns Hopkins University Press.

Turnock, D. 1982. *The historical geography of Scotland since 1707.* Cambridge: Cambridge University Press.

Urry, J. 1981. Localities, regions and social class. *Int. J. Urb. Reg. Res.* **5**, 455–74.

Urwin, D. W. 1977. *The alchemy of delayed nationalism.* Bergen: University of Bergen.

Urwin, D. W. 1980. Towards the nationalization of British politics? The party system, 1885–1940. In *Wählerbewegung in der Europäischen Geschichte,* O. Busch (ed.). Berlin: Colloquium.

Urwin, D. W. 1982a. Territorial structures and political developments in the United Kingdom. In *The politics of territorial identity: studies in European regionalism,* S. Rokkan & D. W. Urwin (eds.). Beverly Hills, Calif.: Sage.

Urwin, D. W. 1982b. Germany: from geographical expression to regional accommodation. In *The politics of territorial identity: studies in European regionalism,* S. Rokkan & D. W. Urwin (eds.). Beverly Hills, Calif.: Sage.

Urwin, D. W. & F. Aarebrot 1981. The socio-geographic correlates of left voting in Weimar Germany, 1924–1932. In *Mobilisation, center–periphery structures and nation-building,* P. Torsvik (ed.). Oslo: Universitetsforlaget.

Vedel, G. (ed.) 1962. *La Depolitisation, mythe ou realite?* Paris: Armand Colin, Cah. Fond. Nat. Sci. Pol. 120.

Verba, S. & N. H. Nie 1972. *Participation in America: political democracy and social equality*. New York: Harper & Row.

Viner, J. 1960. The intellectual history of laissez-faire. *J. Law Econ.* **3**, 45–69.

Vivarelli, R. 1979. Revolution and reaction in Italy, 1918–1922. *J. Ital. Hist.* **1**, 235–63.

Vuillemin, J. 1984. *Nécessité ou contingence?* Paris: Minuit.

Wade, I. O. 1977. *The structure and form of the French enlightenment*. Princeton, N.J.: Princeton University Press.

Wagstaff, H. R. 1972. The rural economy of Scotland in a period of rapid decline in agricultural population. *Scot. Agric. Econ.* **22**, 103–12.

Wald, K. 1983. *Crosses on the ballot: patterns of British voter alignment since 1885*. Princeton, N.J.: Princeton University Press.

Walker, M. 1977. *The National Front*. London: Fontana.

Walker, R. 1978. The transformation of urban structure in the nineteenth century and the beginnings of suburbanization. In *Urbanization and conflict in market societies*, K. R. Cox (ed.). Chicago: Maaroufa.

Walker, R. 1985. Class, division of labour and employment in space. In *Social relations and spatial structures*, D. Gregory & J. Urry (eds.). London: Macmillan.

Walker, W. 1980. *Juteopolis: Dundee and its textile workers, 1885–1923*. Edinburgh: Scottish Academic Press.

Wallerstein, I. 1974. *The modern world-system: capitalist agriculture and the origins of the European world-economy in the sixteenth century*. New York: Academic Press.

Wallerstein, I. 1978. Annales as resistance. *Rev.* **1**, 5–7.

Wallerstein, I. 1979. *The capitalist world-economy*. Cambridge: Cambridge University Press.

Wallerstein, I. 1983. *Historical capitalism*. New York: Schocken Books.

Ware, A. 1985. *The breakdown of Democratic party organization, 1940–1980*. Oxford: Clarendon Press.

Wattenberg, M. P. 1984. *The decline of American political parties, 1952–1980*. Cambridge, Mass.: Harvard University Press.

Webb, K. 1977. *The growth of nationalism in Scotland*. Glasgow: Molendinar Press.

Webber, M. M. 1964. Urban places and the non-place urban realm. In *Explorations into urban structure*, M. M. Webber *et. al.* (eds.). Philadelphia: University of Pennsylvania Press.

Weber, E. 1976. *Peasants into Frenchmen: the modernization of rural France, 1870–1914*. Stanford, Calif.: Stanford University Press.

Weber, M. 1925. *The theory of social and economic organization*. 1964 edn. New York: Free Press.

Weber, M. 1949. Class, status and party. In *From Max Weber*, H. H. Gerth & C. W. Mills (eds.). London: Routledge & Kegan Paul.

Weulersse, G. 1910. *Le mouvement physiocratique en France, de 1756 à 1770*. Paris: Gallimard.

White, J. K. 1983. *The fractured electorate: political parties and social change in southern New England*. Hanover, N.H.: University Press of New England.

Williams, R. 1977. *Marxism and literature*. London: Oxford University Press.

Williams, R. 1980. *Problems in materialism and culture*. London: New Left Books.

Williams, W. A. 1981. *Empire as a way of life: an essay on the causes and character of America's present predicament along with a few thoughts about an alternative*. New York: Oxford University Press.

Williamson, C. 1960. *American suffrage: from property to democracy, 1760–1860*. Princeton, N.J.: Princeton University Press.

Wilson, B. 1976. Scots stoke the Orange fire. *The Observer*, January 18.

Winch, P. 1963. *The idea of a social science*. London: Routledge & Kegan Paul.

Winnick, L. 1966. Place prosperity versus people prosperity: welfare considerations in the geographical redistribution of economic activity. In *Essays in urban land economics*. Los Angeles: University of California, Center for Real Estate Research.

Wirt, F. M., B. Walter, F. F. Rabinowitz & D. R. Hester 1972. *On the city's rim: politics and policy in suburbia*. Lexington, Mass.: D. C. Heath.

Wirth, L. 1938. Urbanism as a way of life. *Am. J. Soc.* **44**, 1–24.

Wolf, E. P. & C. N. Lebeaux 1969. *Change and renewal in an urban community: five case studies of Detroit*. New York: Praeger.

Wolfe, A. 1981. Sociology, liberalism and the radical right. *New Left Rev.* **128**, 3–27.

Wood, E. M. 1981. The separation of the economic and the political in capitalism. *New Left Rev.* **127**, 66–95.

Woolf, S. J. 1984. Towards the history of the origins of statistics: France, 1789–1815. In *State and statistics in France, 1789–1815*, J-C. Perrot & S. J. Woolf. Chur, Switzerland: Harwood.

Wright, E. O. 1978. *Class, crisis and the state*. London: New Left Books.

Wright, G. C. 1977. Contextual models of electoral behavior: the Southern Wallace vote. *Am. Pol. Sci. Rev.* **71**, 477–96.

Zaret, D. 1980. Social science without history: from Weber to Parsons and Schutz. *Am. J. Soc.* **85**, 1180–201.

Zimmerman, C. 1938. *The changing community*. New York: Harper.

Zunz, O. 1982. *The changing face of inequality: urbanization, industrial development and immigrants in Detroit, 1880–1920*. Chicago: University of Chicago Press.

Author Index

Numbers in italics refer to text illustrations.

Subject Index

Numbers in italics refer to text illustrations.

Printed and bound by CPI Group (UK) Ltd, Croydon, CR0 4YY

22/10/2024

01777621-0012